T0143045

Sovereignty Blockchain 1.0

Lian Yuming

Sovereignty Blockchain 1.0

Orderly Internet and Community with a Shared Future for Humanity

Lian Yuming
Key Laboratory of Big Data Strategy
Guiyang Innovation-Driven Development
Strategy Research Institute
Guiyang, Guizhou, China

Key Research Project of Key Laboratory of Big Data Strategy
Key Research Project of Beijing Key Laboratory of Urban Science Research Based on Big Data
Publishing Fund Project of Think Tank Program of Beijing Cosmopolis Cultural Exchange Foundation

ISBN 978-981-16-0759-2   ISBN 978-981-16-0757-8   (eBook)
https://doi.org/10.1007/978-981-16-0757-8

Jointly published with Zhejiang University Press

This Springer imprint is published by the registered company Springer Nature Singapore Pte Ltd.
The registered company address is: 152 Beach Road, #21-01/04 Gateway East, Singapore 189721, Singapore

Support From

International Business School, Zhejiang University
Academy of Internet Finance, Zhejiang University

Academic Support From
Institute of Big Data, Zhejiang University of Key Laboratory of Big Data Strategy

Editorial Committee

General Counsel: Chen Gang

Director: Zhao Deming

Executive Deputy Director: Chen Yan

Deputy Directors: Xu Hao, Liu Benli, Lian Yuming

Academic Counsels: Ben Shenglin, Yang Xiaohu, Li Youxing, Zhao Jun, Zhang Ruidong, Zheng Xiaolin, Chen Zongshi, Yang Lihong

Editor-in-Chief: Lian Yuming

Deputy Editors-in-Chief: Long Rongyuan, Zhang Longxiang

Core Researchers: Lian Yuming, Zhu Yinghui, Song Qing, Wu Jianzhong, Zhang Tao, Long Rongyuan, Song Xixian, Zhang Longxiang, Zou Tao, Chen Wei, Shen Xudong, Yang Lu, Yang Zhou

Academic Secretaries: Li Ruixiang, Long Wanling, Li Hongxia

Senior Translator: Xu Xueying

Translators: Xu Xueying, Zhao Baiqi, Jiao Jingyi, Mo Feifei, Guo Mingfei

Translation Reviewers: Xu Xueying, Long Wanling, Guo Guoliang, Xi Jinting

Preface

Humans have just stepped over the threshold of the third decade of a new millennium. It seems that the world steers into an "anchor-free environment" and faces an unprecedented "disorder dilemma". The only thing we know for sure is "the uncertainty of the world". As Nassim Nicholas Taleb wrote in his *The Black Swan: The Impact of The Highly Improbable*, history and society are not advancing slowly, but jumping from one fault to another. Coincidentally, human civilization is often changed by a few "black swan" events. In a world where science and technology are dramatically transforming the world, we have to study the outcomes before all those happen. As Taleb stressed in his book *Antifragile: Things That Gain from Disorder*, we need to build our own "Antifragile" mechanism in order to live a better life in this uncertain era. This "Antifragile" mechanism is the power to overcome volatility and uncertainty, with its emphasis on promoting a Community with a Shared Future for Humanity. Our research convinces that the Orderly Internet and the Sovereignty blockchain are the new keys to a Community with a Shared Future for Humanity.

A Community with a Shared Future for Humanity is a Hyperledger based on the common interests and shared values of humans. Facing volatility and uncertainty, the world takes on the characteristics of "risk society" outlined by Ulrich Beck, a German sociologist. "With higher technological capacity, it is difficult to predict the consequences of technological development, so global technical risks and institutional risks arise." Global deficit has been a serious challenge for mankind. While global threats, e.g. nuclear war, cyber war, financial war, biological war and non-sovereignty power still exist, the "Grey Rhino" incidents, such as global warming and clash of civilizations are irresistible, and the disasters posed by AI taking over humans are always haunting. Yuval Noah Harari reviewed human civilization in his *Homo Deus: A Brief History of Tomorrow* that three major challenges—famine, plague and war—were posing threats the humans no matter where they live. Though the challenges are being addressed in this new century, the conclusion still remains to be discussed as they are still haunting mankind today. Facing the global threats to our survival and progress, humans have common interests in a community with a Shared Future. In addressing the global issues humans face and achieving sustainable development, governments and international organizations will reach a certain consensus and a shared value, i.e. playing an active role in global governance and promoting a Community with a Shared Future for Humanity. The blockchain based on institutional arrangement and governance system is a Hyperledger, reflecting the common interests and shared values of humanity, and that is sovereignty blockchain. Its impact on human society is making those nations divided by land and nationality the new organizations or group based on consensus, and the "digital world" thus formed is blurring the boundary between virtuality and reality. Global governance is proposed in the context of global information revolution and increasing global issues. It pinpoints that there is an irreversible political process, in which authority is disintegrating, thus a global governance system composed of more and more authority hubs is created. The global governance system built on the premise of weakening national sovereignty and blurring boundaries makes sovereignty transfer a kind of existence in global governance practice.

The Orderly Internet aims at building a new social trust relationship. Human life needs order. Whether order exists and to which degree it is realized are important yardsticks to social civilization. Human civilization is a collaborative network based on trust and consensus. Trust is a kind of order, which needs to be obeyed and observed; however, obedience and observance also need to be supported by trust on a deeper layer. In the digital age, digit (algorithm) is the greatest common divisor of global civilization and the basis for the most consensus that humans reach. In light of a series of global governance headaches caused by "trust deficit", humans are in bad need of a new social trust relationship. The sovereignty blockchain will transfer the Internet from Information Internet and Internet of Value to Orderly Internet. From the perspective of a Community with a Shared Future for Humanity, this is a global community of trust. We are witnessing major changes unprecedented in our world. We are bombed by questions in these major changes, e.g. what are changing; to which direction they are changing and finally what the changes will be. These questions are uncertain and unpredictable. However, it is certain that two forces are shaping our lives and then the world. One is digital currency, which will trigger comprehensive reforms in economy. The second is digital identity, which will reconstruct the governance mode of whole society. Driven by digital currency and digital identity, a credible and immutable consensus and co-governance mechanism is to be set up with the aid of Hyperledger, smart contract and cross-chain technology. With programming and coding, this mechanism helps construct a digital trust system. When digital currency and digital identity meet and match the blockchain perfectly, it marks we have entered a new world where network is our computer. With the help of digital network and terminal, the blockchain reconstructs the joint governance pattern of the nation, government, market and citizen. A new era of digital civilization based on digital trust and digital order is on the horizon.

From the spirit of contract to the rule of conscience. The blockchain is important in that it transforms human dependence on humans and objects into human dependence on data. Data becomes a logical starting point of blockchain and the core of governance technology. No modernization of national governance, no modernization of data governance. The digital, cyber-connected and intelligent governance technology has become the key factor in modernizing national governance. For the past half-century, enterprising sci-tech companies were the drives for human civilization.

However, for a long time in the future, the Governance Tech for Social Good will be a major guarantee for human civilization. Tech for Social Good is a signpost to a general, inclusive and universal digital society, which shapes the first feature of the digital society—kindness and altruism. Conscience is the connotation of Tech for Social Good. Wang Yangming's Philosophy, disseminating and popular all over the world, has become one cultural source for a Community with a Shared Future for Humanity. "Yangming Philosophy, as an ethical, moral and philosophical system, ought to be a force for management, governance, decision-making and promotion. It should be a science of management and governance." As Prof. Chung-Ying Cheng, a tenured professor at Department of Philosophy, the University of Hawaii said, the moral idealism with moral conscience as its core in Wang Yangming's Philosophy is undoubtedly a panacea for the inhuman world clad with moral degradation, venality and excessive material desires. The unscrupulous plunder and destruction of nature by humans imbalanced humans and nature, humans and themselves, humans and the world. The problem may become more serious in twenty-first century, so it is necessary to review and retrospect with "conscience", restrain selfish desires through "the extension of conscience", remain true to the original aspiration, and get along with the uncertainty and the turbulent world with the philosophy that "all things are of one unity". This is the essence of Wang Yangming's Philosophy. The statement by Du Weiming "twenty-first century is Wang Yangming's century" means "twenty-first century is a century calling for conscience".

The world has blown a "warning whistle". The events the sci-fi writer Liu Cixin depicted in *The Wandering Earth* are actually happening on earth now. "At first, nobody took notice of this disaster. It was just a wildfire, a drought, the extinction of a species or the disappearance of a city until this disaster is closely related to everyone". Humans are experiencing a period of uncertainty that may be endless and dangerous. The crisis like this is not the first and will not be the last in history. As Klaus Schwab, founder of the World Economic Forum said, we need more forward-looking global cooperation. Only with an initiative of a Community with a Shared Future for Humanity can we resolve the common challenges like imbalanced development, governance dilemma, digital divide, biological insecurity and clash of civilizations. With the vision and foresight of the shared future of humanity, we can jointly build a new global architecture, open up a new governance realm and create a bright future.

Guiyang, China
March 2020

Lian Yuming
Director of Key Laboratory of Big Data Strategy

Introduction

I

A rampant pandemic named COVID-19 broke out as a "black swan" incident and has risen to the world's top concern in the early spring of 2020. The horror of COVID-19 did not lie in its mortality rate, but in its high transmission rate and strong destruction effect. It not only had a devastating effect on the medical order of a nation, but posed a challenge to national epidemic prevention and control mechanism and public health emergency management, an examination on national governance system and capacity. The global outbreak of COVID-19 has triggered our reflections on human destiny. Humans only have one planet, and all nations live in one world. The international community is becoming a community with a shared future in which all the entities are interconnected. In the face of mounting global challenges and a disorderly, out-of-control world order without trust, not a single nation or organization can cope with all the problems on its own. Never before in history that international community is in dire need of a balanced world order.

As President Xi Jinping remarked at the opening of the 2017 Annual Meeting of the World Economic Forum in Davos, "We live in a paradoxical world today. On the one hand, human civilization has advanced to the highest level in history with accumulating material wealth and rapidly advancing science and technology. On the other hand, the world is facing mounting uncertainty due to frequent regional conflicts and global challenges including terrorism and refugee flows, as well as poverty, unemployment, and widening income disparity." On the following day in a speech at the United Nations headquarters in Geneva, President Xi put forth the question of our era, "What's happening to the world and what should we do about it". More than two years later at the closing ceremony of the Sino-French Forum on Global Governance, he stressed again, "The world today is facing great changes unprecedented in the century. While peace and development remain the themes of the times, humans are facing many common challenges, e.g. instability and uncertainty." In recent years in particular, the Grey Rhinoceros and Black Swan incidents, e.g. nuclear war, cyber war, financial war, biological war and the rise of non-sovereign forces, are emerging; new threats caused by atomic, biological, chemical and digital weapons are looming larger; "deficit crisis" such as deficits in governance, trust, peace and development is in need of immediate resolutions. All these are detrimental to the natural rights, security and development of humans.

Humans share the aspiration to build a clean and beautiful world of lasting peace, universal security, common prosperity, openness and inclusiveness. Constructing a Community with a Shared Future for Humanity and a vibrant, inclusive, equitable and harmonious consensus-based world order constructed, governed and shared by all is the best choice to address global issues and challenges, as well as the optimal route to tackle a world order without trust, order and control. The initiative of such a community cannot do without institutional design and technological support. Sovereignty blockchain, a governance technology from governance of technology to governance of institution, is a smart institutional system based on the Internet order with consensus, constructed, governed and shared by all, and is a major governance instrument integrating sci-tech and institutional innovations. A Community with a Shared Future for Humanity based on sovereignty blockchain, essentially a governance community, is of great significance for promoting multilateral cooperation, bettering global governance system and constructing a balanced world order.

II

President Xi Jinping stated, "Sci-tech innovation and institutional innovation should work in a coordinated manner. The two wheels should turn together.[1] As the two basic forms of innovation in human society, sci-tech innovation and institutional innovation, once combined, will exert unprecedentedly powerful impact. Blockchain, an integrated application of peer-to-peer network, cryptography, consensus mechanisms, Smart Contract and other technologies, is hailed as the "the greatest technological innovation of human in the early twenty-first century." It has reconfigured the governance model for the entire social sphere and provided a reliable channel for the transfer and exchange of information and value in the irregular, insecure and unstable Internet. It also delivers a new computational paradigm and collaborative model to build trust with a low cost in an untrustworthy, unreliable and uncontrollable competitive environment. Such a unique trust-building mechanism transfers penetrating supervision and cascade trust."

Blockchain, a cryptography-based distributed database technology, empowers massive distributed computations to support data-intensive computing such as big data mining and analysis. As an advanced form, sovereignty blockchain, an institutional innovation based on sci-tech innovation, is the core of the next generation of blockchain. Simply put, sovereignty blockchain is a technology governance that helps set up an innovative hybrid technology architecture. Based on that, sovereignty blockchain, highlighting legal regulations, is a combination blow of

---

[1] Xi Jinping. *Striving to Build a World Power in Science and Technology*. People's Publishing House, 2016, p.14.

supervision and governance of technological and legal rules, taking into account both the feasibility of technical rules and the authority of legal rules. For this reason, sovereignty blockchain is a rule of technology under legal rules. It aims to resolve the data ownership of nations, organizations and individuals, thus innovate a governance system shifting from consensus structure to co-governance one and co-sharing one. Different from the blockchain with purely data-centric characteristics, sovereignty blockchain, highlighting human subjectivity, serves as a powerful booster to governance modernization with governance technology.

The leap from blockchain to sovereignty blockchain is not just a supplement, but more a contributor of new concepts, new ideas and new regulations to cyberspace governance. The Internet is a complex interaction and open connection of big data in a virtual space that is unbounded, priceless and disorderly. It shifts from everyone-to-everyone information transmission to information exchange and finally to order share. The Internet is also evolving from information internet to value internet and to order internet. Such an evolution from low to high, from simple to complex is actually transforming uncopiable data into copiable data, which, in essence, is the manifestation of a human-centric data stream in the virtual space. The borderless and scalable nature of such a state makes it difficult to confirm, price, trade, trace and regulate data. In a sense, the Internet leaves us in disorders and chaos. However, the birth of blockchain has brought new opportunities as its application breaks the previous disorder, chaos, and insecurity of the Internet and attempts to build a more orderly, secure, and stable new world. In particular, the invention of sovereignty blockchain has incorporated the law into the blockchain system to drive the change from rule of technology to rule of system. By doing so, it accelerates the blockchain governance and system via controlling the uncopiable data flow of the Internet within a regulable and shareable framework.

III

The theories of data sovereignty, social trust and smart contract form the theoretical basis of sovereignty blockchain. The shift from data to data rights is a natural result as human society ushers in an era of digital civilization, in which we not only possess human rights and property rights, but also enjoy data rights. Data rights allow everyone to share data to maximize value, which is an institutional binding for sovereignty blockchain. Trust, a paramount cornerstone to sustain a virtuous social cycle, serves as a cultural binding for sovereignty blockchain. Its underlying governance logic is the trinity of "self-trust + trust from others + trust in others" based on the consensus mechanism in the society of acquaintances, strangers and Internet. Smart contract, a rule of technology brought by contract codification, is a regulatory framework for a credible digital economy, a programmable society, and a traceable government, so it serves as the technical binding for sovereignty blockchain.

Data Sovereignty Theory. As a fundamental and strategic national resource, data is exerting growing impact on economic development, social governance, national management and human lives. Therefore, any unlawful interference in data may infringe on a nation's core interests. Required by national security, citizen privacy, government enforcement and industrial development, data sovereignty has emerged and become a critical concern for individuals, enterprises, and government agencies. The core of data sovereignty lies in attribution identification, which is divided into personal data sovereignty, enterprise data sovereignty, and national data sovereignty. In practice, the fierce game of data sovereignty between various subjects has created a disorderly state putting data into a security dilemma, while the overlapped interest in data sovereignty has triggered problems such as unclear data ownership and chaotic data circulation and utilization. In order to ensure data security and maintain data sovereignty, each entity should build data sovereignty under the premise of holding a holistic view of national security, handle the relationship between sovereignty and human rights properly, and advance the harmonious development of data sovereignty and data human rights for people's data welfare.

Social Trust Theory. Trust is the fundamental source of individual sense of security and the inner foundation of interpersonal relationships and social operation. As the concept of humans evolves in history, the construction of order also requires a change in the type of trust. Blockchain, supported by technology, has established a set of decentralized, open and transparent trust system, where trust subjects and objects can establish mutual trust relying solely on an algorithm-driven distributed network. Such a system can be built without the need to get the creditworthiness of the other party or to endorse the creditworthiness of a third party. As such, a self-trust society that does not need credit accumulation comes into being and makes "data equaling credit" a reality. The thread "digital currency—digital living—digital life—digital economy—digital society" runs through future development, and a community with shared digital trust is the cornerstone. In the ideal digital society of the future, there will be abundant material wealth; issues of trust and order will gradually be resolved, "humans" will reach a state of free and conscious existence, and society as a whole will be a federation of digital citizens.

Theory of Smart Contract. From identity to contract, from agrarian civilization to industrial civilization and then to digital civilization nowadays, humans are stepping from "an era of human rights" and "an era of property rights" into "an era of data rights". Accordingly, the law undergoes a dramatic transformation from "the law of people" to "the law of objects" to "the law of data". As a set of protocols, smart contracts can automate certain tasks that need to be performed manually otherwise. The existing legal system, embedded into it, provides an effective method to adapt law to the rapid development of new technologies. Decentralized, transparent, credible and tamper-resistant blockchain provides a reliable, fair and impartial implementation environment for smart contract. With the deep integration of big data, artificial intelligence, blockchain and other new-generation information technologies, a digital twin virtual world will be generated on the physical basis, in which the information and intelligence can be delivered and exchanged in a digital world, between human and human, human and objects, or between objects and objects. Humans will enter a smart society where every object is intelligently connected. Such brand-new intelligent social relationship established through blockchain, especially through smart contract, will serve as a pillar cornerstone for a new order of digital civilization.

IV

Global issues should be resolved by global governance. A Community with a Shared Future for Humanity is a Chinese approach to the future of the world, human development and global governance. This global governance mechanism highlights the equality and common values of humanity, namely, seeking common grounds while reserving differences and living in peace. In such a context, nations will follow common rules and mutual trust, thereby dramatically reducing the cost of governance in human society and empowering humans to create maximum value while consuming least resource. As a buzz word in international relationship and global governance, a Community with a Shared Future for Humanity, repeatedly mentioned and elaborated by Chinese leaders in multiple speeches at home and abroad, has been included in a number of resolutions of United Nations agencies and become a global consensus to a large extent. In March 2018, "The initiative of a Community with a Shared Future for Humanity" was written into the preamble of the Constitution after the vote on the first session of the 13th National People's Congress. The initiative was included in the Constitution and the legal system of the PRC, which means this initiative is now a main component of "rule of China".

The proposition of a Community with a Shared Future for Humanity is both reality-based and future-oriented. However, it is difficult to achieve in the near future due to the existence of selfish departmentalism, protectionism, and insufficient consensus and trust mechanism in regions, nations and ethnic groups. Humans trapped in Prisoner's Dilemma are not opt to choose local optimum compared with global optimum. To this day, human society has always been operating in local optimum, but a Community with a Shared Future for Humanity proposed by China is a mode that facilitates human society to achieve a global optimum, for it integrates China's worldview of "harmony sustains mutual trust and mutual assistance among nations" and its governance concept of "harmony in diversity". Therefore, the cornerstone of such a community lies in the establishment of a low-cost, high-credible and ultra-secure trust mechanism and governance mode.

History shows it is hard to establish full trust simply through institutional design. Reliable trust-building techniques are also needed as a foundation. Blockchain precisely solves this problem, but meanwhile, it cannot replace the social organization in the exercise of institutional functions and cannot embody the attributes of a social system, the sovereignty will of a nation, or the values, ethical preferences and cultural traits of a society. To put it another way, blockchain faces a huge challenge in terms of the institutional function of human society. The only solution is to change the exercise subject of social institution from social organization to certain technical approach. Sovereignty blockchain is exactly a perfect choice. It aims at harnessing legal regulation under the premise of national sovereignty and building different levels of intelligent institutional system according to different data ownership, functional positioning, application scenarios and open access, with distributed ledgers as its basis and rules and consensus as its core. The sovereignty blockchain-based Community with a Shared Future for Humanity is oriented at the value of a global governance in which everyone is responsible, everyone does his part and enjoys the benefits, thus forming the common code of conduct and value norms for human society and promoting a real era of Orderly Internet globally.

Contents

1 **Orderly Internet** 1
  1.1 The Evolution of the Internet 1
     1.1.1 Internet of Information 2
     1.1.2 Internet of Value 6
     1.1.3 Orderly Internet 9
  1.2 Relationship Between Data Force and Data 12
     1.2.1 Data Evolution 13
     1.2.2 Data Capital Theory 18
  1.3 Data Game Theory 20
  1.4 Sovereignty Blockchain 23
  1.5 From Blockchain to Sovereignty Blockchain 24
  1.6 From Unitary Dominance to Decentralized Co-governance 30
  1.7 From Dependence on Objects to Dependence on Data 32
  References 34

2 **Data Sovereignty Theory** 37
  2.1 Data Rights 37
  2.2 Human Rights, Property Rights, and Data Rights 38
     2.2.1 From Numbers to Data Rights 38
     2.2.2 Definition of Data Rights 39
     2.2.3 Distinctions Between Human Rights, Property
        Rights, and Data Rights 42
  2.3 Private Rights, Public Power, and Sovereignty 44
     2.3.1 Private Rights 44
     2.3.2 Public Power 46
     2.3.3 View of Sovereignty 47
  2.4 Sharing Rights: Essence of Data Rights 48
     2.4.1 Sharing Rights and Possession Rights 48
     2.4.2 From "One Ownership for One Object"
        to "Multi-Ownership of Data" 49
     2.4.3 Connotation of Sharing Rights 50
  2.5 Data Sovereignty and Digital Human Rights 51

2.6   Differences Between Sovereignty and Human Rights 51
        2.6.1   Tension and Conflict Between Sovereignty
                and Human Rights 52
        2.6.2   Communication and Agreement Between
                Sovereignty and Human Rights 53
2.7   Human Rights of the Fourth Generation 54
        2.7.1   The First Generation of Human Rights 54
        2.7.2   The Second Generation of Human Rights 57
        2.7.3   The Third Generation of Human Rights 58
        2.7.4   Digital Human Rights 59
2.8   Blockchain Reshapes Data Sovereignty 60
        2.8.1   Re-Understanding Data Sovereignty 61
        2.8.2   Sovereignty Under Blockchain 62
        2.8.3   Data Sovereignty Under Blockchain 63
2.9   Data Sovereignty Game 65
2.10   Confrontation of Data Sovereignty 65
2.11   Transfer and Sharing of Data Sovereignty 67
2.12   Data Sovereignty, Data Security and Data Governance 70
        2.12.1 Challenges and Countermeasures of Data
                Sovereignty 70
        2.12.2 The Essence of Data Security: National Data
                Sovereignty 71
        2.12.3 Data Governance Rises to National Strategy 73
References 74

3   Social Trust Theory 79
    3.1   Trust and Consensus 79
    3.2   Trust and Social Order 80
            3.2.1   Trust and Reliance in an Acquaintance Society 80
            3.2.2   Trust System in a Stranger Society 81
            3.2.3   Trust Crisis in a Network Society 83
    3.3   The Consensus Mechanism of Blockchain 84
    3.4   Trust is De-Trust 87
            3.4.1   Centralization and Decentralization 87
            3.4.2   De-Trust of Blockchain 88
            3.4.3   Blockchain-Based Credit Society 90
    3.5   Digital Trust Model 92
    3.6   Model Theory and Trust Model 92
    3.7   Hyperledger: From Traditional Bookkeeping
            to Everyone-To-Everyone Bookkeeping 95
    3.8   Cross-Link Transmission: From Information Transmission
            to Value Transmission 96
    3.9   Smart Contract: From Choice Trust to Machine Trust 98
    3.10   Digital Currency and Digital Identity 101
    3.11   Digital Currency 101

3.11.1 Hazards of Digital Currencies 101
3.11.2 Digital Currencies and Sovereignty Digital
Currencies 107
3.11.3 Sovereignty Digital Currency Under Sovereignty
Blockchain 113
3.12 Digital Identity 115
3.12.1 From Traditional Identity to Digital Identity 115
3.12.2 Digital Identity and Destratification 117
3.12.3 Blockchain-Based Digital Identity 119
3.13 Digital Order 121
References 125

4 Theory of Smart Contract 129
4.1 Trust Digital Economy 129
4.1.1 Blockchain Empowers Financial Technology
(FinTech) 130
4.1.2 Smart Contract and Data Transaction 134
4.1.3 Trusted Digital Economy Ecosystem 137
4.2 Programmable Society 142
4.2.1 From Being Programmable to Cross-Terminal 142
4.2.2 Data, Algorithm, and Scenario 146
4.2.3 Digital Citizens and Social Governance 149
4.3 Traceable Government 152
4.3.1 The Regulation of Government Power 153
4.3.2 Traceability of Government Responsibility 156
4.3.3 Paradigm of Digital Government 160
References 164

5 Community with a Shared Future for Humanity Based
on Sovereignty Blockchain 167
5.1 Human Destiny in Global Challenges 167
5.1.1 Future Global Challenges 168
5.1.2 Risks Behind "Grey Rhino" and "Black Swan" 171
5.1.3 The Gospel or Disaster of Artificial Intelligence 175
5.2 Governance Technology and a Community with a Shared
Future for Cyberspace 179
5.2.1 Governance Technology and Governance
Modernization 180
5.2.2 Three Pillars of Governance Technology 184
5.2.3 Cyberspace Governance Based on Sovereignty
Blockchain 192
5.3 Tech for Social Good and Rule of Conscience 199
5.3.1 The Value Orientation of "Data Persons" 200
5.3.2 The Soul of Science 203

       5.3.3   Cultural Significance of Yangming Philosophy
              in Building a Community with a Shared Future
              for Humanity 209
    References 217

Postscript 223

Glossary 227

Bibliography 231

Chapter 1
Orderly Internet

We stand on the threshold of a brave new world. It is an exciting and if precarious place to be.

—Steven Hawking, Famous British Physicist.

The Internet is the first thing that humanity has built that humanity doesn't understand, the largest experiment in anarchy that we have ever had.

—Eric Schmidt, Former Google Chairman and CEO.

You don't need to know much about blockchain technology, just like you don't need to know much about Internet technology.

—Jack Ma, Founder of Alibaba Group.

1.1 The Evolution of the Internet

The emergence of Internet[1] and the technology iteration which is brought by it bring a new human language, a new concept of thinking, and a new human civilization, and humans enter a brand-new era. From the Internet of Information to the Internet of Value and to Orderly Internet is the fundamental rule for the Internet to evolve from being basic to advanced. Internet of Information resolved information asymmetry and enabled humans to acquire the dividend of convenient communication and

---

[1]Lao Zi, an ancient Chinese sage, profoundly expounded the development rules of the interaction and interconnectivity, mutual symbiosis, and mutual prosperity of all things in the *Tao Te Ching*. We can use Lao Zi's ideological framework of "One *Tao* gives birth to one, one to two, two to three, and three to all things" to interpret the Internet. The way of the Internet is freedom, and the pursuit of freedom has spawned the desire for self to connect with others freely. This is the "one" of the Internet. The decentralized structure of the Internet guarantees independence in its connection with everyone. Everyone becomes the center and everyone is me; meanwhile, everyone is centered on others and I am everyone because of decentralization. This unity of opposites "Everyone is me; I am all" constitutes the "two" of the Tai Chi structure of the Internet. Binary interaction has spawned the "three" elements of the Internet: people, information, and transactions. The dynamic integration of the three elements gave birth to the Internet miracle in various forms. (Tang Bin. The Internet is the Romance of a Group of People. *Chinese Business*, No. 5, 2015, pp. 122–123.).

© Zhejiang University Press 2021
L. Yuming, *Sovereignty Blockchain 1.0*,
https://doi.org/10.1007/978-981-16-0757-8_1

information cost reduction. With the development of E-commerce, the Internet has enabled the transmission of value as convenient, fast, and low-cost as the transmission of information; in particular, with the development of blockchain, humans came to realize the potential of data asset appreciation and value system reconstruction. While with Orderly Internet, humans came to realize the prospects of the innovative organization mode, governance system, and operation rules of sovereignty blockchain. The development of Internet, in three stages, is actually a process from traditional internet to intelligent internet and from informational technology to digital technology. In the future, the fusion of blockchain and Internet will reconstruct new cyberspace, impacting humans immeasurably.

1.1.1 Internet of Information

Every information revolution, from language to characters, and from the invention of printing to Internet boom, made a revolutionary effect on human society. Two events which happened in 1969 made the year to be remembered. One was man's first moon landing and the other was the birth of the Internet. The former meant man began to explore the universe, while the Internet served to connect different computers, both of which represented the network development from one-point existence to multipoint existence and the extension of humans (Chen 2015). In 1993, the emergence of World Wide Web made a democratic internet. In 1995, the publication of *Being Digital* announced the digital migration of human society. If the last Age of Exploration expanded human physical space, the age of exploration then expanded human digital space. From the theory of evolution, the 50-year-old Internet is still quite young, but "the Internet has become such a fixed fixture in our minds that it's even easier to imagine the end of our lives than to imagine life beyond the Internet" (Winter and Ono 2018). Kevin Kelly even believed that if there is extraterrestrial life, they will invent electricity, lights, and cars, and eventually the internet. Just as Eric Schmidt said, "the Internet is the first thing that humanity has built that humanity doesn't understand, the largest experiment in anarchy that we have ever had". The internet brought humans into an infinite, borderless, and endless status, penetrated into every corner of modern civilization free of charge, transboundary, open, democratic, and with long-tail effect and multiple values. Therefore, the resulting internet world has created more surprises than those brought about by railways, telegraphs, and other inventions ever.

Transitions of Information. With the extension of medium and the progress of carrying mode, the speed, breadth, and dimension of information transmission is also undergoing unprecedented changes. According to Metcalfe's Law, as devices and users join, the value and importance of the network will grow exponentially.[2] The

---

[2]The network must be of a huge number: the first phone is useless; the second is a bit useful, but it is limited to talking with the first phone; then there are thousands of phones, making sense to buy one; after millions of phones, a phone will really become an indispensable tool.

invention of telephones is the first node, which facilitated one-on-one information transmission; the invention of radio and television is the second, facilitating one-to-N transmission; while the invention of Internet is an important node for realizing the information transmission efficiency to get to N^2. The Internet achieved information interconnection through TCP/IP protocol, which opened an era of information explosion. Netizens worldwide can transmit and receive information without distinction through the Internet. The information transmission efficiency extends from "1 to N" to "N^2". With the Internet, everyone has "microphone" and "Moments". Everyone can independently set up issues and transmit discourse, and the self-empowerment capacity and autonomous orientation which cannot be endowed by real system and physical space came into being (Changshan 2016). The internet unshackled the confines of time and space, the boundary between virtuality and reality. The digital and physical boundaries are disappearing, and humans are heading to a borderless society. The Internet provides an informal and virtual social space structure for human society, in which any government, organization, or individual can participate and shape us into a new species—"data persons", just like "economic persons" in classical economics. As described in the documentary *The Internet Age,* "the great migration of human life has begun either worldwide or in China, which is a comprehensive migration from traditional society to the digital age of internet, and is an epoch-making subject and unstoppable human destiny. You are in this great migration whether you are a netizen or not and whether you are far from internet or immerse in it". Information intercommunication is realized with the internet, interpersonal communication with mobile internet, and after the emergence of 5G, the Internet of Things (IOT) will take the "fast lane" to transit from the Internet of Everything to Artificial Intelligence + Internet of Everything. The internet eliminated the barriers to information transmission, allowing us to enter the Internet of Information era of free information transmission. Internet of Information refers to the internet which records and transmits information, and information we mention here is reproducible and can be reproduced at a low cost. The Internet of Information is revolutionary in that it not only impacts the traditional game logic and rules but creates infinitely extended value foundation and space.

Mutual Trust in Internet of Information. While enjoying all the benefits of information explosion, humans also suffer from the pains brought by the Internet, endure the chaos and disorder of network world, and even feel the risks and fears brought by the network. Since "there is no unified and pre-set beat in the internet, everyone dances freely according to their minds", (Yongchao and Qiping 2012) the internet becomes a tool for new violence, a part of a new weapon, and the foundation of new power.[3] We enjoy convenience and swiftness in the business empire of Internet

---

[3]Random search introduces to many related terms or cases, such as digital coercion, digital violence, cyber-bullying, cyber-violence, and forced digital participation, coerced digital participation, etc., digital burglary, digital fraud, digital extortion, and digital robbery are often more serious than traditional theft, fraud, extortion, and robbery.

of Information created by BAT,[4] transmit our chat information, shopping information, and transaction information on these huge central agencies, and meanwhile have to accept the disadvantages brought by this double-edged sword, including lack of trust, leak of privacy, information plague, business monopoly, internet polarization, online scams, and hacking. The degree of "disorder" and uncertainty of the Internet is steadily growing, and the Internet governance as an epochal proposition demonstrates "disorder". According to Chaos Theory, "the original state of everything is a seemingly unrelated pile of fragments, but when this chaotic state is over, these inorganic fragments will be organically gathered into a whole". The Internet amplifies the opposition between information and noise, and the chaos of the Internet of Information brings about information inefficiency, overload, and distortion. Information invalidity is mainly manifested as insufficient information acquisition around specific problems or multi-party problems, which then leads to unconscious community behaviors, resulting in the dependence of decision-making on aftereffects. Information overload is the most serious problem on the Internet of Information era. If insufficient information may bring risks to supervision and decision-making, information overload will increase the cost of selecting supervision and decision-making information. The huge amount of information brings a surge of information while the uncertainty of its analysis is also increasing. Uncertainty is positively correlated with the degree of social disorder, and the higher the uncertainty is, the more serious the social disorder will be. Information distortion means the Internet is different from traditional media in that it lacks an effective information quality control mechanism. Therefore, the real information on the Internet is often easily out of the control of an originator and owner and distorted in transmission. The Internet realizes the "decentralized" dissemination of information; however, there are two fatal problems. One is related to consistency, while the other is to correctness. This is Byzantine Failures, which refer to "synchronization" and "trust" in decentralized information dissemination. "The order directed to freedom is not necessarily the home of freedom, but the yoke of freedom" (Shuying 2009). To some extent, the Internet of Information is in a state of chaos and disorder, and there are uncertainty and disorder risk in the Internet space.

Internet of Information: Borderless, Priceless and Disorderly. When the Internet is no longer uncopiable, humans are intoxicated by the happiness of the free transmission of information and they have to face the confusion caused by the borderless, priceless, and disorderly Internet of Information. The Internet of Information is first borderless. It is infinite and has no boundaries. Then it is priceless. The Internet, just like air, has value but no price, so it cannot be priced. Third, it is disorderly as it is chaotic. "As Steve Jobs, who was deeply influenced by the hippie spirit, said, 'The computer is the most extraordinary tool that man has ever created. It is like the bicycle of our mind.' The bicycle, as a tool of vagrancy and rebellion, gives one freedom to reach a destination without a track. The Internet, growing up in the embryo of computer, is a chaotic world with the flag of freedom flying everywhere"

---

[4]China's three major Internet companies—Baidu (B), Alibaba (A), Tencent (T)—the initials of their names.

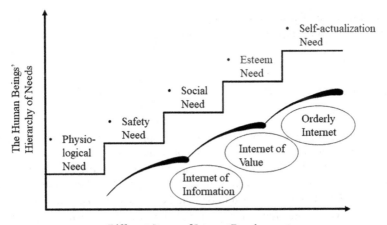

Fig. 1.1 Maslow's demand model of the Internet

(Xiaobo 2017). The disorder of the Internet is innate and it is directly related to the borderless and priceless, which is the biggest trouble brought about by the Internet. The Internet is running in the borderless wild like a wild horse, and the consequences will be horrible if there is no rein on the horse. To be a better horse, we need to highlight order and rules to address the connection, operation, and transformation of the Internet. Humans can quickly generate and transmit information to every corner of the net world via the internet, but they have been unable to solve the problem of value transfer[5] and credit transfer. In short, the internet addresses information asymmetry successfully, but it fails to address value asymmetry, or credit storage, therefore. The evolution of the network follows the path of "growth → breakpoint → balance". First, the network will grow exponentially; then, it reaches a breakpoint when it has grown beyond its capacity and its storage must decrease (slightly or significantly); finally, it will reach a balance and its quality rather than quantity will grow sensibly (Stibel 2015). "Maslow's Hierarchy of Needs" divides human needs into five levels. We extended this theory to the Internet needs to establish a hierarchical model of Internet needs (Fig. 1.1). [6] At present, there are growing demands in human society and more urgent needs for order innate in human nature. On the one hand, the boundary of Internet of Information and Internet of Value will continue to

---

[5] Value transfer means, if we want to transfer part of the value from A to B, A should explicitly lose this part of the value, and B explicitly gains this part of the value. This operation must be approved by both A and B, and the result cannot be controlled by either A or B. The current Internet protocol fails to support this practice, so the transfer of value requires the endorsement of a third party. For example, the transfer of A's money to B via the Internet often requires the credit endorsement of a third party.

[6] From the perspective of satisfying people's demands, information Internet satisfies the human social needs to expand the scope of social contact, the Internet of Value satisfies the human respect needs to gain recognition of value, and the Orderly Internet satisfies the highest human needs of self-realization.

extend with the revolution of technology. On the other hand, higher level demands of humans, e.g., for trust and order, are growing.

1.1.2   Internet of Value

With the progress of technology, the Internet has stepped into great discovery of value, great creation of value, and great innovation of value. Blockchain, as the cornerstone of the Internet of Value, has the inherent capacity to convey trust and value, and to reconstruct rules and order. According to the White paper "Realizing the Potential of Blockchain" released by the World Economic Forum, the blockchain will usher in a more subversive and transformational era of the Internet, generate new opportunities, promote the creation and transaction of social value, and shift the Internet from the Internet of Information to the Internet of value. The Internet, with its emergence and popularization, makes it extremely easy for people to establish point-to-point connections on the internet. Compared with the Internet which makes information transmission easy, blockchain appears in the form of a completely open data block information chain, realizing point-to-point transmission and exchange of value, thus becoming a new engine of the internet and starting the era of Internet of Value.

Interconnectivity of Value. The Internet of Information has realized information transmission from being closed, delayed, cumbersome to fast, free, and convenient, but information exchange is not up to people's increasing demand for value. The birth of blockchain has unveiled new dawn for the evolution of the Internet of Information. Value transmission and exchange can be realized through the Internet of Value based on the blockchain protocol. The Internet of Value means people can transfer value, especially assets, like transferring information on the Internet without any third-party intermediary or medium (Fig. 1.2). Internet of Value is the real

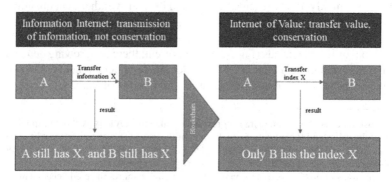

Fig. 1.2 Information internet and internet of value

Table 1.1 Comparison between Information internet and internet of value

Project	Information internet	Internet of value
Function	Information transfer	Value transfer
Mode	Copy	Record
Representation	Link	Pass
Protocol	World wide web protocol	Blockchain credit layer

embodiment and transparent transfer in which internet value forms value interconnectivity chain and realizes Internet value based on blockchain protocol, aimed at the interconnectivity of capital, contract, data, trusted identity, and other values (Table 1.1). Although the Internet makes it possible to transfer funds and contracts, these functions rely on the centralized institutions to transfer information to corresponding centralized institutions; however, they bring about high transaction costs, information asymmetry, passive acceptance, and privacy insecurity. Why blockchain leads the Internet of Value is that it resolves the above difficulties in exchange, transaction, and transfer, so people can transfer value on the Internet as convenient, fast, safe, reliable, and low-cost as transferring information. With blockchain, people can transfer capital and data assets to every corner of the world like sending WeChat messages and Microblog. The relationship between Internet of Value and Internet of Information is not a substitute relationship but to solve problems based on different application scenarios. Internet of Value superimposes value attributes on the basis of the Internet of Information, thus a new type of Internet to transmit information and value comes into being.

Blockchain: A Major Cornerstone of Internet of Value. The blockchain has brought new development space for the Internet and triggered a new stage of development. Blockchain, a transition from the Network of Information to the Network of Value, helps promote the transmission of mutual trust-based value, overturn business model, break industrial pattern, and change distribution system. Blockchain is characterized by decentralization, tamper-resistance, retention of traces throughout the process, traceability, collective maintenance, openness, and transparency, etc. The Internet is in a state of disorder, monopoly, chaos, and insecurity; however, in the Internet of Value, blockchain facilitates natural trust between people and puts an end to the centralized monopoly of the traditional Internet. Counterfeit-resistance and tamper-resistance of blockchain enable every person to establish his/her own integrity node in the network. Under the dual supervision of system and technology, a person will be doubly punished by the legal system and Smart Contract once he/she commits a crime, and people will then take maintaining credit as a habit in a subtle way. It must be pointed out that the transmission of both information and value is ultimately the transmission of data. The credibility of data is the foundation of the Internet of Value, and data ownership confirmation is the foundation of data credibility. Only trusted data is valuable in computing and analysis and in providing intelligent services later on in order to realize business contract, contract datalization, and data trust. The trusted infrastructure of Internet of Value is the most important thing for value exchange and application and industrial ecology. Blockchain is the core technology for data confirmation, value exchange, and benefit payment, so it lays the foundation for the Internet of Value via data confirmation and exchange. The first is the real and sole data conformation; the premise of value is to determine the ownership of assets. The sole ownership of assets is guaranteed with cryptography as well as the public and private key mechanism. The consensus mechanism guarantees a chronological sequence of claiming ownership and the first one who claims ownership is the true sole owner of asset. Distributed ledgers ensure permanent and immutable historical ownership. The second is safe and reliable exchange. Supply and demand show value and there is no value without exchange. With cryptography, owners release their assets to others by providing signature verification. The consensus mechanism prioritizes trades, resolves the "double-spending" of assets,[7] and records confirmed trades. Smart Contract guarantee that real transactions only happen automatically if they meet certain conditions.[8]

Governance: Biggest Challenge of Blockchain. The Internet is not only a new object of national governance but a new means, tool, and platform of governance. In the Fourth Plenary Session of 19th Central Committee of the Communist Party of China, it was stressed that "strengthening and innovating social governance, improving the system of social governance with the CPC committee leadership, government responsibility, democratic consultation, social collaboration, public participation, legal guarantee, science and technology support"; the "science and technology support" was added to modify "social governance system" for the first time, which embodies full and highly conscious confidence in developing, utilizing, and managing Internet and other technology. As the Internet of Value becomes a global infrastructure like the Internet of Information, Smart Contract, as a decentralized network protocol that is automatic, open, and transparent, will ensure the rules of the Internet of Value are credibly enforced and will usher in a new era of contracts. The governance of the Internet of Value is much more complex than that of the Internet of Information, as the network that makes up the Internet of Information is based on a unified global platform, while the blockchain is an account made up of different and sometimes rival accounts. The dilemma is "Mundellian Trilemma", meaning you cannot have scalability, distribution, and security at the same time. To exchange values between stakeholders with global peer-to-peer resources, blockchain needs a set of reasonable and orderly governance frameworks to ensure the effective play

---

[7]Don Tapscott, "father of digital economy", stated in a speech, "The last few decades have ushered in the information age of the Internet. When I send you an email, a ppt file, or something else, I'm not actually sending you the original, I'm sending you a copy. But when it comes to assets, such as money, stocks and bonds, membership points, intellectual property, music, art, votes, carbon credits, and other assets, sending copies is not a good idea. If I give you $100, what's important to me is that I don't have it anymore, and I can't send it to you again. This is what cryptographers call the 'double consumption' problem, or 'double sending'. By creating Bitcoin, Nakamoto solved the double sending of digital certificates. Furthermore, Internet of Value needs to solve the double sending of all digital assets on the Internet.

[8]Yi Huanhuan. *The Value of the Internet and Blockchain: Four in One New Network*. Sohu.com, 2018, https://www.sohu.com/a/249565405_100112552.

of its technical potential. According to the White paper "Realizing the Potential of Blockchain" released by the World Economic Forum, challenges in the eco-governance of blockchain include the lack of a reasonable legal governance structure, immature legislation or norms that may hinder the progress of blockchain, the speed of blockchain in its application exceeding that of technology in maturity, lack of diversity of ideas, and the monopoly of authorities of the entire network, etc. In addition, there are some unknown challenges, e.g., waste of resources, multiple complications caused by innovative integration, possible reinforced regulation, terrorist incidents, attacks by quantum computing, and technological failures. For example, though blockchain technology aims to improve trading ability, terrorists may use these advanced technologies to create more and greater chaos once it is released open online. For another example, quantum computing, as a cryptographic algorithm beyond electronic computing, is likely to deal a fatal blow to the technological advantages of blockchain. What's more, if blockchain technology turns out to be buggy, it could lead to the collapse of the entire Internet of Value.[9]

1.1.3   Orderly Internet

In the development of the Internet, many trends and approaches have been questioned, but its core idea—the necessity of the Internet—has not been questioned. The best future for the Internet is not one of immortality or constancy, but rather questioning whether the Internet we possess today is the best we can design, and considering alternatives to human lives when the Internet era is over.[10] Now, it is necessary and urgent to transform the Internet in a broader and deeper way than ever before. The properties of blockchain can precisely respond to this demand, maintain the ecological order of the cyber world, and then construct a more benign governance framework so as to empower the modernization of national governance system and governance capacity effectively. The Internet has become a place for the game of forces, but there is a more urgent call of this place for order and more serious lack of order and responsibility in the real world. Order sustains Internet and the Internet will be destroyed in disorder without rules and order. Internet governance is forming a certain order, but in order formation, it is inevitable to have a game between the government, enterprises, society, and individuals in the borderless Internet as the battlefield. In the face of order shock with the coexistence of deconstruction and reconstruction, there is dynamic power conversion between strong and weak in the game, and data sovereignty theory, social trust theory, and Smart Contract theory constitute the core elements and governance logic of the Internet order. Louis Kahn, a renowned American architect, once said, "The world will never need Beethoven's

---

[9]World Economic Forum. "Realizing the Potential of Blockchain", CCID Translation. 2017, No. 47, pp.1–19.

[10][US] Jennifer Winter, [Japan] Ryota Ono. The Future of the Internet, translated by Zheng Changqing. Publishing House of Electronics Industry, 2018, p. 39.

Fifth Symphony until Beethoven composed it. Now we can't live without it". In the age of the Orderly Internet, we are faced with more unknowns, and if there's only one thing we know, it's that we'll create more things people can't live without.

Three Inferences for the Future. The first basic inference is that human society is moving from a binary world system to a ternary system. Humans used to live in a world composed of physical space (P) and human social space (H), and the order of their activities was formed by the interaction and interplay between human and human, between humans and objects. Humans make and lead human social order. Cyber, digital, and intelligent technologies break through the physical space and reconstruct it digitally, and the information space (C) becomes a new pole of the world space (Yunhe 2018). In this new space, data is the breeding ground for everything. The world is transforming from traditional binary world (P, H) to ternary (C, P, H) world; the order of human activities is apt to be reconstructed accordingly. Laws of production and life, forms of social organization, social governance system, and legal system norm, which are formed and operated based on the original binary world, will face challenges and reshaping brought about by the development logic of ternary world, and be in bad need of the theory and practice. The second basic inference is that the era is moving from oil-driven to data-driven. At a higher stage of economic and social development, the development path driven mainly by factors and investment is no longer sustainable. Victor Mayer-Schonberger, the father of big data, believed that "sometimes it is not necessarily the idea that drives the change of the world. It may be the real data that generates ideas on the basis of data. New ideas are the core of creative destruction, while data is the driving force of innovation". Innovation drives development and data drives innovation. Data is oxygen, a key resource to transform the world and the core element of order construction, bringing new balance to mankind. The third is that the Orderly Internet is the future of the Internet. As the center of all human activities, it combines technical rules with legal rules to realize the sharing of credit and order, which is an advanced form in the development of the Internet. As the most revolutionary technology of the twentieth century, the Orderly Internet may be the leading force for future innovation-driven development. It is of great significance to improve the global governance system of the Internet and the initiative of A Community with a Shared Future for Cyberspace.

Three Forces to Shape the Future. Three major forces are shaping the world in ways that have never been imagined before. The first is data rights. "Information is the center of power", (Weigend 2016) and social power under the Orderly Internet becomes a systematic force. Futurist Alvin Toffler said that, as a power over others, rights have been acquired through violence, wealth, and knowledge since ancient times. Data empowerment makes social power shift from violence, wealth, and knowledge to data rights. The data rights, the product, and inevitable trend of the times when humans move toward digital civilization are a major force in the reconstruction of order. This power marks the decline of traditional power, the expansion of new power, and the transfer of individual sovereignty. The second force is altruism. "Humans are moving from the era of IT (Internet Technology) to the era of data technology". The relationship structure of the data technology era determines its internal mechanism is decentralized, delayering, and borderless, and its basic spirit is openness, sharing, cooperation, and mutual benefit. These characteristics determine the "people-oriented" humanistic background of the society and the core value of "altruism" in this era. Altruism and sharing are the key forces to a new round of sci-tech revolution and industrial transformation. Based on altruism and sharing, human civilization is destined to move to a higher stage and to an order constructed by sharing rights. The third force is remixing. Remixing is the rearrangement and reuse of existing objects, the fusion of internal and external resources, and the creation of new value. Growth comes from remixing, including the progress of civilization, economy, and data... Remixing is a crucial subversive way, an inevitable and natural force for changes. The value proposition of data rights, the value orientation of altruism, and the value innovation of remixing are the essential elements of the Orderly Internet, and they are shaping this world with their interplay.

Three Judgments for the Future. The first is that the rise of China is inevitable. No force can hold back the progress of the Chinese people and the Chinese nation. The cardinal impetus for China's rise lies in deepening reforms and opening up in a comprehensive way. Any emerging country has gone through a special historical stage of meeting increasing risks and challenges in their growth from being big to being strong. The unique advantages of Chinese institution are our confidence in confronting risks and challenges. From a longer timeline, the rise of a major nation is bound to go through obstacles and the key is to maintain strategic resolve and stamina and focus on managing our own affairs well. "The wind extinguishes a candle, but it inflames a fire". The second basic judgment is that the real hallmark of China's rise is the modernization of national governance and the establishment of an international discourse power in the global governance system. As China is standing in the historic intersection of "Two Centenary Goals", a new round of technological revolution and industrial revolution becomes a major power in shaping the world pattern and the rise and fall of big power. The reshaping of global economic governance system has made it possible for China to get a more favorable international status. The rising instability of international political pattern, the strategic containment of America on China, in particular, will be the largest uncertainty in China's external environment. What China will face in the future is not only a contest of hard power such as economy, sci-tech, and military strength, but more importantly and symbolically, a competition of comprehensive soft power such as national governance system and governance capability. The rise of China will bring about great changes in the pattern and mode of global governance. The third judgment is that governance technology is the core force of national governance modernization. Governance technologies marked by artificial intelligence, quantum information, mobile communications, the Internet of Things, and blockchain impact the international landscape in three ways. First, they increase global wealth significantly. Second, they reinforce economic and military changes and directly affect and change the balance of power among nations. Third, the power of non-state actors is growing rapidly, even posing new challenges to ideology, security, and global strategic stability. The Orderly Internet will become a key factor in the modernization of national governance. Taking advantage of this critical strategic opportunity, China will promote infrastructure, legal system, and standard system based on the Orderly Internet, so as to seize the institutional voice in global governance.

The future is more uncertain but more promising. "The evolution of the Internet is chaotic at the micro level, but it shows surprising direction at the macro level. Like the 'invisible hand' in economics, business activities are disorderly at the micro level, but there is a balancing force at the macro level. The evolution of the Internet is more magical than economics as it is not a balancing act but a monotonic increase" (Feng 2012). The Internet is a highway leading to the future; big data is the vehicle driving on the highway, and the blockchain is the system and rules that protect vehicles to drive legally and orderly. The Internet is an irregular, insecure and unstable world, while blockchain makes the world more orderly, safer, and more stable. If the Information Internet resolves the borderless issue, the Internet of Value the priceless issue, the Orderly Internet the disorder issue of the Internet. Based on the Orderly Internet, we will usher in a new digital planet in which individuals, enterprises, and nations have to awaken themselves from old experiences to keep up with the changes of the times and successfully "immigrate" to the new planet.

1.2 Relationship Between Data Force and Data

With the steady evolution of the Internet, data grows in an explosive way. Blockchain, with its trustworthiness, security, and immutability, allows more data to be freed and truly circulated. Big data makes a smarter world. "Worldwide of data, all social relations can be represented by data, and humans are the sum of relevant data" (Mayer-Schonberger and Cukier 2013). Productivity and production relationship are the most important concepts in human society. In the era of big data, the relationship between data power and data exists undoubtedly. In a general sense, data drives the era of big data, and this force brands the production relationship with data relationship. The relationship between humans and technology, humans and economy, and humans and society are therefore facing unprecedented deconstruction and reconstruction. Human society is at a critical turning point in history: old balance and old order are gradually disintegrating, and a new system and order are emerging. From the three-sided game to the integration of the trinity in the ternary world, it is necessary to review the new relationship between humans and technology, humans and economy, humans and society in order to reconstruct the relationship between humans and the world. Humans will certainly usher in a revolution. "Although we can't name the revolution, but as long as it aims to bring the world into final stability and balance, it must be a revolution for the harmonious coexistence of humans and nature, society and society and human and human, to create social systems, values, and ways of living" (Munesuke 2007). In the era of Orderly Internet, the data philosophy based on data evolution theory, data capital theory, and data game theory with human origin indicates the growth of civilization and the reconstruction of order.

1.2.1 Data Evolution

Evolution, a biological theory, is scientific evidence of the origin and development of species. Taking data as the core element, "Data evolution" reviews and studies the relationship between humans and technology and its essential laws from the perspective of historical materialism and dialectical materialism. It is the technology that determines the future best. "Datalization" is significant in that "it is a technology that is completely different from industrial technology and can create infinite value without consuming a lot of limited resources. It is a technology field that can maintain infinite happiness and move people constantly (Munesuke 2007). In this sense, datalization is a technical field that can make the society as a higher level "stable and balanced system" possible under existing conditions. In a certain sense, human existence is also a kind of technological existence. Thus, a new era of humans emerges in which new technological structures prop new social structures. Human society gradually evolves into a "tech society". "Technology and coastal engineering have enabled about 10 million Dutch people to live below sea level behind levees, without which there would have been no Dutch people", Bijack (2004) expressed by an American scholar Vibe E. Bijak on the importance of levee technology for the Dutch. In a larger horizon, it can be said there would be no human today without technology. Humans come into being, grow, live and evolve in technological activities. If humans are the sum of relevant data, then the relationship between humans and technology is the essence of technology.

Key Technologies Impacting the Future World. Human society is in a new technological revolution marked by an "integration" revolution. Its essence is "remixing", and the core is to create unprecedented "super machines" via the "integration" of various data technologies (Table 1.2). "History has shown us that major technological changes lead to social and economic paradigm shifts" (Zarkadakis 2017). The theme of *What Technology Wants*, Kevin Kelly's best-selling book, is that technology is independent of humans and technology dictates the course of the world. The latter sentence is true in a way, just as mankind is about to face what may be the last upgrade—thorough technicalization after the advent of agriculture, the development of industry, and capitalism. The technologies that have caught our attention in recent years, e.g., 5G, gene editing, blockchain, edge computing, quantum information, have brought about the speculation and imagination of their subverting the era and even civilization. "Technology, as a social activity, is a form of human self-expression, embodying the purpose and value of humans. The humanistic development of technology should follow the people-oriented principle and focus on human survival and free development" (Ning and Shujun 2018). Technology has given humans a lot of positive meanings, but humans, unable to get away from the "GE-Stell" power of technology, visualize their own species in danger in the technology era. "Technology dilemma" and "value nihilism" are the destiny that humans cannot avoid (South Reviews Office 2019). Even Heidegger said tactfully both "yes" and "no" to technology, but he quoted a line from Hölderlin expressing the belief that "where there is danger, there is salvation". The storage, transmission, and computing capabilities

Table 1.2 Technologies impacting the future

New technologies	Dimension of Impact	Impact	Feasibility	Impact score
AI	Human daily life	Many jobs will be replaced, and traditional industries will embrace new forms of business and development	High	5
Internet of Things		Intelligent home life and social security system; development of Industrial Internet	High	5
Virtual reality, augmented reality		The fusion of real environment and equipment; more successes in major medical operations	High	4
4D print		Customized Home Innovation factory; Cancer Vaccine	Middle	3
Robots	Human mode of production	E-commerce platforms deploy robots in order implementation, warehousing and distribution; Intelligent home butlers; robot rescue and disaster relief, etc	High	5
Blockchain		Data security and privacy protection in the financial sector; supervision and protection of insurance business	High	5
New energy		Improvement of environment; energy diversification	High	5
Brain-machine interface	Human capacity	To bring light to the blind; rehabilitation of mobility impaired persons; Human IQ increment	Middle	4

(continued)

Table 1.2 (continued)

New technologies	Dimension of Impact	Impact	Feasibility	Impact score
Gene sequencing		Genetic examination; prevention and treatment of cancer; Treatment of genetic diseases	Middle	4
Quantum technology	Technology	Communication technology with absolute security; Very efficient quantum computing	Middle	5
Terahertz		New applications in mobile broadband communications, anti-stealth radar, counter-terrorism, non-destructive industrial testing, food safety testing, medical and biological imaging	Middle	5

of information drive human civilization. Meanwhile, human civilization also faces an insurmountable gap, which is the channel that restricts the rapid improvement of information storage, transmission, and computing capacity, namely, the mobile communication technology and key information infrastructure from wired to wireless, among which 5G technology is the most fundamental. 5G technology will bring about revolutionary changes that will affect human society in the future. The changes are characterized by high speed, ubiquitous network, low power consumption, low delay, Internet of Everything, and reconstruction of the security system. Its essence is to upgrade the storage, transmission, and computing capacity of human information, and boost humans from industrial civilization to digital civilization. 4G changes life, 5G changes society, and 6G innovates the world. According to the tempo of "developing one generation of products while constructing one, constructing one generation of products while using one" in the mobile communication industry, it is expected that 6G will be commercially viable around 2030. At present, the Finnish government is the first worldwide to launch a large 6G research program. The Federal Communications Commission of the United States has opened the Terahertz spectrum for 6G research, and China also started to research 6G in 2018. 6G will create a new technological ecology with new architecture and new capabilities as well as the new needs and new scenes of social development, and propel human society to a digital twin world integrating virtuality and reality (China Mobile Research Institute CMRI 2019).

A Multidimensional Review of Human-technology Relationship. Human evolution is closely related to the development of technology. Technology drives human development and increases the possibilities for humans. "Every technology tends to create a new human environment", asserted McLuhan (Spinello 1999). The domination and expansion of intelligent network-based digital waves have penetrated into every corner of human society. Negroponte has a famous saying in *Digital Survival*, "Computing is no longer only about computing; it determines our survival" (Negroponte 1996). Technology constitutes the basic situation of humans and sets a new framework for human survival. The human-technology relationship is steadily evolving.

At the stage of manual skills, humans were combined with technology and there was more dependence of technology on humans. As a primitive technology, manual skills belonged to the human body, something innate in the human body, and they were integrated with the human body and could not exist independently of the human body. Humans were alienated from technology at the stage of machine technology. The relationship between humans and technology was fundamentally changed by machines, and the value of humans has been replaced by that of machines. Technology developed from human technology to instrumental technology, from physiological technology to mechanical technology, and from human dependence technology to separable technology (Dehong 2004). It boosted humans from natural existence to artificial existence, which enhanced human technologization, led to the dislocation of technology-human relationship, and the alienation of human technologization; therefore, technology came to deviate from or even challenge human nature. Humans competed against technology at the stage of smart technology. Smart technology is not a continuation of modern technology, but a new stage of technology development, which endows technology with new essence. If mechanical technology is a technology that represses human nature, smart technology is a technology that calls for human nature (Dehong 2003). The upcoming smart civilization will be characterized by the objects invented by humans to devour or control humans, which indicates that the human brain may be surpassed and conquered by cloud brain and super brain. Kevin Kelly named technology the seventh existence of living things. He believed humans and technology can coexist, and the evolution of technology is to the benefits of humans. Humans and technology, as well as life and human creation, are constantly co-evolving in an "infinite game", rather than winning and losing in a "zero-sum game".

Natural Persons, Robots, and Gene-edited Men. Humans have developed to the eve of great changes. Yuval Noah Harari stated in *Homo Deus: A Brief History of Tomorrow*, that science and technology, represented by big data and artificial intelligence, are increasingly mature, and humans will face the biggest change since the evolution to Homo sapiens.[11] Digitalization is not only a technical system, the bit conversion of everything, but the reorganization of human production and life style,

---

[11] From the birth of life on earth 4 billion years ago to the present day, the evolution of life follows the most basic laws of natural evolution, and all life forms change in the organic field. But now, for the first time, it is possible to change this pattern of life.

as well as an updating social system. More importantly, this kind of updating may even reconstruct human social life (Zeqi 2016). Data, a part of our lives and even our life, is profoundly changing the image, connotation, and extension of "humans". With technological revolution, the functions of "natural persons" are gradually deteriorating. Or we may say, "The physical functions of natural persons degraded enough, and the delivery of intellectual functions to robots is under way. The ego degrades as much as the natural person delivers. In this process, it is possible for a robot to replace a natural person as the protagonist of human society" (Caihong 2016). If robots, merely "integrating" human functions, surpass humans, then the precise manipulation of the human genome on viable embryos through gene editing could create artificial "gene-edited man". Gene-edited man does not have "congenital deficiency", whose physical strength and intelligence foundation are superior to those of natural persons. With the advance of biotechnology, it is easy for a man to "integrate" the information of history, ethics, culture, and so on. With the innate powerful ability, gene-edited man is undoubtedly better than the natural person in an all-round way. In *The Selfish Gene*, the author describes it this way, "We are all survival machines–robots as vehicles, programmed randomly to perpetuate the endowed selfish factors called genes". The implication is that human organism, as a machine, leaves much to be improved. One-dimensional technological progress is insufficient to create more welfare for humans. The logic of information transmission and social progress lies in the interaction between humans and technology.[12] There is no end to technology, no end to evolution. In the near future, human society is likely to be composed of natural persons, robots, and gene-edited men, and the conflicts between them are likely to exceed the current conflicts among nationalities, religions, and civilizations. The future may be a marriage of the organic world and the synthetic world.

For decades, successive waves of technology and the disruptive changes that followed have consigned established standards to history. Technology boosted human civilization and social productive forces and causes the long-term opposition and separation between nature and technology. Currently, smart technology, biotechnology, and other technologies propel humans to evolve from natural persons to robots and gene-edited men, thus they enter a new "post-Darwinian evolutionary stage" or "post-human" stage. Gene-edited men may have been born if ethical constraints were removed. For now, artificial intelligence is only overtaking humans in certain areas, such as computing power, and it is no more terrifying than cars running faster than man. The real "singularity" to watch out for is when machines become self-aware, or even self-replicating (Kotler 2016). But in the face of unprecedented advances in technology, which new technologies will drive major changes in humanity? What will the world be? I'm afraid everyone wants to know the answers to these questions.

---

[12]As Kasting stated, having a computer doesn't necessarily change the world. It's the use of computers by people that change the world. Information network is not always a good thing, as the network has no emotion; it can either serve humans or destroy humans and everything depends on the program people prepare, so it is a social and cultural process.

1.2.2 Data Capital Theory

If the theme of the society in eighteenth and nineteenth centuries was "machine", which reflected the understanding and application of man toward various substances in nature, then we are stepping into an age of data in the twenty-first century. Data is becoming the core asset of this era. As the main factor of production, creation, and consumption, it affects and changes all aspects of society, especially the organizational form and value creation of companies.[13] In his book *Data Crush*, Christopher Surdak defined data capitalism as the "singularity" in the history of capitalism. He said that "We will move from a world which takes capital as the basis of wealth and power to one which takes data as the basis of wealth and power... Over the next 15 years, the world will shift its focus from capital to data" (Sanhu 2017). Data growth leads to economic growth, and data is disrupting the old economic order. The proposition of "data capital theory" provides a new theoretical explanation and multiple dynamic choices for the world economy as it studies the essential relationship and internal law between humans and economy in the new era.

Changes in Data Value. The era of big data is marked by the fact that data has become basic social resources and the elements of economic activities. In addition, it is an inevitable trend for the development of big data to make data resources, assets, and capital. Data is a new factor of production. The Fourth Plenary Session of 19th Central Committee of the Communist Party of China proposed to "improve the mechanism whereby labor, capital, land, knowledge, technology, management, data and other factors of production are assessed by the market and remunerated according to their contributions". This was the first time that the central Chinese government proposed the distribution of data according to the contribution as a factor of production. As a factor of production, data participation in distribution can be regarded as a continuation of the logic and development of technology participation in distribution from a certain perspective, so it is of far-reaching significance. "We are entering the age of data capital". Guo Yike, director of the Institute of Data Science at Imperial College London, summarized the development of data economy into four stages. The "day before yesterday" of data, that is, data information stage. In the past, data was only a material for recording and measuring the physical world. The data "yesterday", that is, at the data production stage, when data was used to provide a service, it became a resource, a product, and thus a series of data products and services were born. The data "Today", that is, at the stage of data assets, people have realized the definition of data ownership made it an asset and the basis for wealth generation and data started to become an essential part of the total assets of individuals. The data "tomorrow", namely, at the stage of data capital, where data assets connect the value. Data assets realize their value through circulation and transaction and finally become capital. Technological progress is manifested

---

[13]Tian Suning. *Continuing Along the Road of Knowledge*. [Austria] Victor Mayer-Schonberger, [Germany] Thomas Ramsey. *Das Digital*, translated by Li Xiaoxia, Zhou Tao. China CITIC Press, 2018, preface 1.

in the improvement of data processing, analysis, and application associated with data capital accumulation, which is the prerequisite for data to become capital. "This process also transforms data assets into data capital that can directly drive productivity development, thus creating a production function different from previous industrial revolutions: productivity = (data-processing capacity) labor + data capital + data capital performance technological progress" (Jianfeng 2014).

From Ownership to Right of Use. In industrial economy, the right of control and right of use within ownership are integrated (Qiping 2012). In the digital age, the ownership (in effect, control in ownership) and the right of use are being separated.[14] In the future, the right to use will be more important than the ownership. It is better to use rather than possess. The essence of use is to open up one's own resources to exchange and connect with others. "Global economy is moving away from the physical world towards the non-physical bitcoin world. Meanwhile, it is moving away from ownership towards the right of use; it's moving away from the value of replication toward the value of the network; And to a world that is bound to come, where remixing is growing and happening" (Kelly 2016). At present, there is a widespread practice of separation of ownership and right of use. Although the legal structure of data ownership is still being studied, facts show that data ownership does not matter. What matters is who has the right to use the data and what value the data can produce. The key to data property is the separation of ownership and the right of use, which is transforming the old economic order. Data is non-depleting, replicable, shareable, severable, non-excludable, and of zero marginal cost, etc. On the one hand, data is a special commodity with value and use value; on the other hand, it is more a kind of expandable capital. With this characteristic, digital labor becomes the source and carrier of value in the era of big data. The basic law of data movement has promoted the depth and breadth of global value chain reconstruction and brought about new ways of competition and growth. The power of data has brought about a profound change in the relationship between data, which is triggering a broader economic and social movement to shift a competitive economy to sharing economy. Sharing is an irresistible and transformative force. In the future, more and more social resources will begin to be shared. The essence of sharing economy is to "weaken ownership and release the right of use". Sharing rights makes the separation of data ownership and use right possible, and forms a sharing development pattern of "not owning, but using". Sharing value theory will surely become a revolutionary major theory after the theory of surplus value.

From Efficiency to Fairness. After the publication of *Capital in the Twenty-First Century* by Thomas Piketty, a French economist, the imbalance of global wealth

---

[14] As early as 2000, Jeremy Rifkin wrote, "The shift from the idea of property rights to the idea of sharing is a structural shift away from the market and the transaction of property rights and towards the ideologically driven interpersonal relationship. For many people today, this transformation is as inconceivable as enclosure, the privatization of land, and labor became a property relationship five hundred years ago. A quarter of a century from now, the concept of ownership will be decidedly limited and even somewhat anachronistic for a growing number of enterprises and consumers. [US] Jeremy Rifkin, *The Zero Marginal Cost Society*, translated by CCID Institute expert group, CITIC Press, 2017, p. 241.

distribution has attracted wide attention. Every technological revolution in human history has pushed forward the progress of human society to different degrees, and meanwhile, it has also brought about the imbalance of power and wealth between different groups of humans.[15] The age of data capital, in fact, means a technological revolution and a revolution in business models, but different from other revolutions in history, this one comes to bridge the gap between the rich and the poor worldwide. With the interplay of data force and data relationship, the world has moved from "The Era of Division of Labor" to "The Era of Combined Labor", achieving large-scale social synergy across organizational boundaries. In the remixing world, trans-boundary occurs at any time, and resources in one field recombined with resources in another field lead to new innovations. The digitization of resources will inevitably lead to the reallocation and redistribution of resources. The digitization of distribution will accelerate the unification of efficiency and fairness. This kind of reconfiguration and redistribution is an updating social system, which helps form a new socio-economic model and bridge the digital divide. While rationally allocating resources and pursuing efficiency, such reallocation and redistribution will effectively bridge the gap between the rich and the poor caused by digital exploitation, improve social equity, and promote the dynamic balance between equity and efficiency. The future world should be "delayering", in that rising digital currency is a key channel for transferring money to the poor in the future; emerging digital identity creates development opportunities for global vulnerable groups; the digital governance under construction leads to global management system reforms. Therefore, the gap between the rich and the poor can be bridged; opportunities, identity, and status will be equal, and different groups will have equal access to information, to collaboration, and to free communication.

1.3   Data Game Theory

The Paradox of Massive Data. With the exponential growth of the amount of data, people's understanding of the value of data is increasingly consistent, and data is discovered, unsealed, and mined in large quantities. If we say the socialization of

---

[15] With the invention of the steam engine in the second half of eighteenth century, Europe entered the age of industrial civilization, while many parts of the world were still in agricultural civilization, and the gap between the two in wealth became increasingly obvious. From this point on, the center of global wealth began to shift to the West. With the improvement of relevant systems such as stock trading at the beginning of twentieth century, New York began to be the world's largest financial center, and the New York Stock Exchange, Wall Street and J.P. Morgan Chase began to be symbols of the contemporary financial industry. The financial revolution also widened the gap between America and Europe. With the explosion of innovation in electronics, communications, semiconductors, software, and other fields in the second half of twentieth century, Silicon Valley began to be a mecca for global information industry. After information technology revolution, the gap between the United States and other regions, such as Asia, widened. As American technology was exported to the rest of the world, wealth and power were further concentrated in the West.

science and the scientization of society are the two basic symbols of the century of science, then the coming century is to complete the socialization of data and datalization of society. While humans are delighted to see the value of data, countless garbage data in the society becomes "data puzzle" affecting human data acquisition ability, storage ability, analysis ability, activation ability, and prediction ability. The resulting structural contradiction between data supply and demand, social contradiction between data protection and utilization, antagonistic contradiction between data public right and private right, and the competitive contradiction between strong data and weak data nations would have existed for a long time. Human cognition of data value is roughly divided into three stages: first, the era of small data based on empirical science to judge data value; second, the era of big data that mines data relationships with data resources as factors; third, the era of hyperdata that governs data congestion with the mark of data explosion. In the era of small data, the bigger the data, the higher the value; In the age of hyperdata, the bigger the data, the lower the value. In the era of small data which lacks data, backward data collection, storage, transmission, and processing technology lead to limited access to data, and it is difficult to make rapid, comprehensive, accurate, and effective research and prediction on things through multidimensional data fusion and correlation analysis. In the era of hyperdata overloaded with data, data explosion brings information overload and data deluge, which makes humans wrapped by data garbage layer by layer. These problems and dilemmas are called "data congestion". The characteristics of big data, such as large scale, multiple types, and fast speed, mentioned by Victor Mayer-Schonberger, will become its fatal weakness. Data garbage brings cognitive barriers to humans, and data congestion may be a major issue of global governance in the future.

Game is Governance. In the blockchain world, there is a popular saying "computing power is power; code is law". In addition, one more sentence "game is governance" can be added, i.e., the process of game is the governance process of decentralized self-organization governance. First, strengthening social good governance abilities in the game. With the progress of technology, the governance of network community is more reflected as a distributed and pluralistic co-governance mode with the participation of multiple subjects including the government, Internet enterprises, network organizations, and netizens. Meanwhile, international cyberspace governance also embodies the pattern of multilateral and multi-party participation, with various entities achieving a certain balance and effectiveness in mutual games. The state of non-cooperative game, a basic form of social operation under technological conditions, highlights the inherent requirements of social sharing development brought about by social datalization. The second is to enhance social autonomy in the game. According to Schumpeter's innovation theory, innovation means "establishing a new production function" and introducing a new combination of production factors and conditions into the production system, which is also the function of platforms. In addition, a very important feature of Nash equilibrium is the consistency between belief and choice. In other words, a faith-based choice is rational, and the beliefs that support it are also correct. So, the Nash equilibrium has the self-fulfilling character of a prediction. If everyone thinks it will happen, it will. Mr. Nakamoto said

Bitcoin is a self-fulfilling prophecy. This consistency and self-actualization between belief and choice allow society to function as a perpetual motion machine. The third is to enhance social governance ability in the game. Under the strong control of a single government, an orderly state appears on the surface of the network society, but with stronger government governance, the innovation vitality of the network society begins to weaken. At this time, the participation of governance forces such as society and enterprises help solve the negative issues brought by strong government control. By forming the social co-governance model of the Internet, the network is revitalized; the network society governance tends to be orderly and efficient, and the governance level becomes higher. The future governance network society calls for the participation of more enterprises and social organizations, and they will become the best practitioners of multi-governance and sharing in the network society. The government should be the general coordinator and the general moderator of multiple stakeholders, as well as the general supervisor and the general custodian of public interest. The functions of the government will be more represented by top-level design, overall planning, coordination, rule-making, security guarantee, and social mobilization abilities.

The Balance of Ternary World. The essence of the era of big data is that the world is moving from binary to ternary, and its major task is to optimize ternary balance. The key to this rebalancing is to pay more attention to the use of external resources to consolidate their strategic position and carry out sustainable development. The ancient "heaven, earth and humans", the basic understanding of the composition of the world, can be the embryonic form of the ternary world. In fact, the ternary relationship is the relationship of balance and stability, cooperation and sharing. The world power has the characteristic of "three *yuan* (three elements)", that is, independent power, balanced power, and coexisting power, which are combined into one body, and jointly maintain, balance, and keep the world. It is the joint action of the three powers that drives the universe and the survival, growth, and continuous growth of human society. From balance to imbalance and then to a new balance is the law of the development of things. After the spiral of "balance–imbalance–rebalancing", everything will continue to develop. It is precisely in such a relatively balanced motion that the universe exhibits infinite charm and perfect regularity. The equilibrium motion is relative and dynamic, and once an equilibrium is broken, new conditions will be created to balance it. "The closer you are to the center, the farther you are from failure". In the era of Orderly Internet, a more dynamic balanced vision is more conducive to social development. Kevin Kelly's nine laws of nature for making things out of nothing emphasize the shift from "making things out of nothing" to "making changes out of change". In other words, any large complex system changes in a synergic way, and only in symmetry and equilibrium can a safe center be formed. The world needs to operate around this center in order to achieve dynamic balance in the change so as to avoid risks. Only in dynamic balance the curse of the life cycle is broken, sustainability and rebalancing achieved, and a new life cycle of future global governance established.

Why Information Grows, by American scholar Cesar Hidalgo, is hailed as "an important milestone in the theory of economic growth in twenty-first century", which proposes a theme: the essence of economic growth is the growth of information or the growth of order. Hidalgo believed that nations good at promoting information will be more prosperous. Data evolution theory, data capital theory, and data game theory (the new "Three Theories") are exactly reconstructing the new order of humans and technology, humans and economy, humans and society in the digital civilization era. In the age of digital civilization, the essence of growth is not the growth of GDP, but the growth of civilization and order. The proposition of the new "Three Theories" reshapes the social structure, economic function, organizational form, and value world, redefines the future human social composition with natural person, robot, and gene-edited man as subjects, and redistributes the new-type rights paradigm and power narrative with data as key elements. In short, the law of data movement reconstructs the order between humans and technology, humans and economy, and humans and society, which is not only a grand idea for the study of future life but a significant discovery for studying the development of digital civilization and the evolution of order.

1.4 Sovereignty Blockchain

Human society is on the eve of the outbreak of new theories and new technologies. Nikola Tesla, a great scientist, remarked that "new technologies are emerging to serve the future, not the present". Blockchain[16] is an integration technology, a data revolution, a reconstruction of order, but a turning point of the era. In this era of changeable order, diffuse rules, and lack of rationality, blockchain has become a leading force for humans to construct order. Blockchain is a transformation of technology, organization, and behavior mode, and reconstruction of governance mode, regulatory mode, and legal rules based on technology. Through extensive consensus and value sharing, blockchain promotes the formation of new value weights and measures of human society in the era of digital civilization and gives birth to new credit system, value system, and rule system. The integration of blockchain and the Internet will technically transform the copiable into the non-copiable or the conditionally copiable. This condition is to move from borderless, priceless, and disorderly to bounded, valuable, and orderly. Sovereignty blockchain provides an alternative path and unlimited imagination for the development from Information Internet, Internet of Value to Orderly Internet. If the blockchain has consensual technical attributes, then the sovereignty blockchain is a unity of consensus, co-governance, and sharing. From

---

[16]McKinsey & Company believes that blockchain is the core technology with the most potential to trigger the fifth wave of disruptive revolution after steam engine, electricity, information technology, and Internet. Just as steam releases people's productivity, electricity satisfying people's most basic living needs, information technology and the Internet completely changing the business model of traditional industries (such as music and publishing), blockchain technology will make it possible to realize the safe decentralized transfer of digital assets.

blockchain to sovereignty blockchain, its significance is not only in the development of blockchain, but in bringing new concepts, new ideas, and new regulations to cyberspace governance.

1.5   From Blockchain to Sovereignty Blockchain

An unidentified person or group, with the name Satoshi Nakamoto, introduced the concept of "Bitcoin" to those on encrypted mailing lists in 2008 and created the Bitcoin social network, developing the first block, the genesis block[17] in 2009. However, others believed the blockchain concept, as a technological innovation, was proposed as a "distributed accounting" system by Stuart Haber and Scott Stometta in 1991. To improve the accuracy of digital documentation, they argued if you didn't trust any single individual or institution", then trust everyone; i.e., let everyone worldwide be a witness to digital documentation". In terms of concept, this idea is indeed similar to that of the Bitcoin blockchain: decentralization is essentially multi-centralization—since there is no center of authority, everyone becomes a center, and each center needs self-discipline as well as other rules, mutual verification, and balances to build a well-knit trust machine (Qian 2019).

Rediscovering the Blockchain. The essence of blockchain is a trust network, which constructs a low-cost mutual-trust mechanism to realize the value transfer of network and Smart Contract recorded in a machine language instead of a legal language (Table 1.3). From the perspective of data, blockchain is a distributed database or distributed ledger that is difficult to be changed. It facilitates the joint maintenance of all participants via a decentralized form. From a technical point of view, blockchain is not a single technology, but a collection of technologies and their results. These technologies combine in new ways to create a new way of recording, storing, and expressing data (Xiaomeng and Shujian 2018). Blockchain possesses key core technologies, such as point-to-point protocol, timestamp, game theory, consensus mechanism, data storage, encryption algorithm, privacy protection, and Smart Contract. It is naturally characterized by multi-party maintenance, cross-validation, whole-network consistency, and tamper-resistance. In comparison with other technologies, its core features are as follows: first, "autonomy". The "autonomy" of blockchain is a challenge to the current Internet organization and architecture; second, "credibility". The credibility of blockchain is a challenge to the current trust architecture of the Internet and even the entire human society. So far, its development can be divided into three stages: Blockchain 1.0, which is represented by

---

[17]Its founding paper Bitcoin: a point-to-point electronic cash system proposes a new decentralized electronic cash system, one of the core ideas is by means of peer-to-peer network, the single centralized dependence is eliminated, point-to-point transactions realized. Meanwhile, the serial number database of the spent digital currency is transformed into the serial number database of the unspent digital currency, and the data scale is controlled. In addition, the hash algorithm is used to mark time and connect them.

Table 1.3 Definitions of "Blockchain" by domestic and foreign organizations

Institutions (organizations)	Definition
UK Government Office (*Distributed Ledger Technology: Beyond Blockchain*)	Blockchain is a database that stores records in a block (rather than collecting them on a single table or paper). Each block is "linked" to the next block with a cryptographic signature shared and collaborated between anyone with sufficient authority
Ministry of Industry and Information Technology of the People's Republic of China (*China Blockchain Technology and Application Development White Paper*)	Blockchain is a chained data structure that combines data blocks in chronological order and connects them sequentially, and a distributed ledger that cannot be tampered with or forged by cryptography
China Blockchain Technology and Industrial Development Forum (China's first Blockchain Standard *Blockchain Reference Framework*)	Blockchain is a non-forgery, tamper-resistant and traceable block data structure built on transparent and trusted rules in a peer-to-peer network environment, which is a mode to achieve and manage transactions
World economic forum	Blockchain (distributed ledger technology), a collection of cryptography, mathematics, software engineering and technology, and the theory of behavioral economics in an emerging technology, uses a global peer-to-peer network of multiple nodes to jointly record data, which ensures the fairness of value exchange between billions of devices without the need for trusted third-party endorsement
International business machines corporation	Blockchain is a shared ledger technology that allows any participant in a business network to view transaction system records (ledger)
KPMG	Blockchain, core technology of Bitcoin, is a decentralized database ledger
McKinsey	Blockchain is essentially a decentralized distributed ledger
Accenture	In Bitcoin transactions, blockchain technology is used as a publicly distributed general ledger for recording transaction information. Multiple blocks share transaction data and records in a point-to-point manner, forming a distributed database
DTZ	A blockchain is a distributed database system that supports and provides continuous transaction records, known as blocks, that cannot be changed or modified. Each block has its own timestamp and is connected to a block that records information about the previous transaction. Blockchain is an important core technology of Bitcoin

(continued)

Table 1.3 (continued)

Institutions (organizations)	Definition
Investopedia	The blockchain is a public ledger that all bitcoin transactions need to be backed by
Wikipedia	Blockchain is a licensed distributed database based on the Bitcoin protocol that maintains a growing list of data records that cannot be tampered with or modified, even for the operators of the database nodes

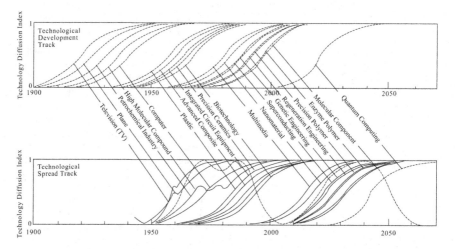

Fig. 1.3 Evolution path of basic innovation in the sixth Kondratieff cycle. *Source* Orient Securities

Bitcoin, and distributed accounting as a typical application; Blockchain 2.0, represented by Ethereum, with Smart Contract technology introduced; Blockchain 3.0, "Blockchain + ", is fully integrated into all aspects of information life (Shen et al. 2019). If history repeats itself, we are at the beginning of the sixth Kondratieff Cycle[18] in human history (Fig. 1.3). The Internet is the core technology construction of the five rounds of Kondratieff cycle since the Industrial Revolution. It has deeply changed the way of human production, life, and survival, and the terminal has almost become an extension of human body functions. At present, the fourth Kondratieff cycle has come to the end; the fifth Kondratieff cycle has turned around, and the sixth Kondratieff cycle may have begun. Basic innovations in the sixth round of Kondratieff cycle may include new materials, artificial intelligence, Internet of Things, genetic engineering, quantum computing, and blockchain, etc. Kaname Akamatsu, a Japanese economist, argued that the world system has a "centre-periphery" structure. Because of the late development advantage, the gap between peripheral nations and central

---

[18] Kondratiyev long wave ("Kang Bo" for short), or Kondratiyev long cycle, is used to describe the long-term fluctuation of economic growth, which usually lasts for 40 to 60 years.

nations in comprehensive national strength shows a "convergence and divergence" cycle, with a time span of 20–60 years. Its long edge corresponds to the long edge of Kondratieff cycle. Blockchain will be one of the core infrastructures of the sixth round of Kondratieff Cycle. Whoever can master the core technology of blockchain will dominate the future world system. If the Internet, big data and artificial intelligence are the ships for humans to enter digital space, blockchain is the sail of that ship. Blockchain will be the underlying technical protocol for human society to govern the digital world. We may probably lose our way in sailing toward the future without blockchain. "Without the Internet, the US might still be the US today, but China is certainly not the China today".[19] So is blockchain, no one will reject the blockchain in the future, as no one can exist without the blockchain.

Sovereignty Blockchain: Core of Next-Generation Blockchain. In the development of global Internet, human society will build a community with a Shared Future for Cyberspace on the premise of respecting national sovereignty behind cyber sovereignty. The sovereign Internet needs to be collocated with a matching sovereignty blockchain. Sovereignty blockchain is to be set up under law and supervision, based on distributed ledger, with rules and consensus as the core to realize the mutual recognition of different participants and thus forming the delivery of public value. In the future, on the basis of sovereignty blockchain, different economies and nodes can realize the circulation, sharing, and appreciation of the consensus value across sovereignty, center, and field, thus forming the common code of conduct and value norms of the Internet society.[20] The foundation of sovereignty blockchain is blockchain, which is a kind of rule of technology that will innovate a set of hybrid technology architectures. Based on this, the sovereignty blockchain is the rule of technology under legal regulation (Fig. 1.4), by highlighting legal regulation and integrating technological innovation and institutional restructuring. The sovereignty blockchain does not aim to be unconventional or unusual, or to overlap or add to the complexity. Its development is in line with the basic law of internal and external interaction. First, the sovereignty blockchain will innovate the modern governance model in a comprehensive way. For the subject of national governance, the sovereignty blockchain can not only realize the iterative renewal of governance approaches, but accelerate the evolution and innovation of governance mechanism, and ultimately promote the new paradigm revolution of modern governance, namely, from closure to openness, from monopoly to sharing, from centralization to decentralization, and from one dimension to multidimension. In terms of government governance, the upgrading of government governance under the sovereignty blockchain will create a brand-new shared, open, collaborative and digital government, and a governance system based on digital technology evolving from consensus structure to co-governance structure and then forming a shared structure. Second, the sovereignty blockchain facilitates the organic integration of people, technology, and society. The

---

[19] Wu Xiaobo. *Tencent Biography (1998-2016): Evolution of Chinese Internet Companies*. Zhejiang University Press, 2017, p. 19.

[20] Guiyang People's Government Information Office. *Guiyang Blockchain Development and Application*. Guizhou People's Publishing House, 2016.

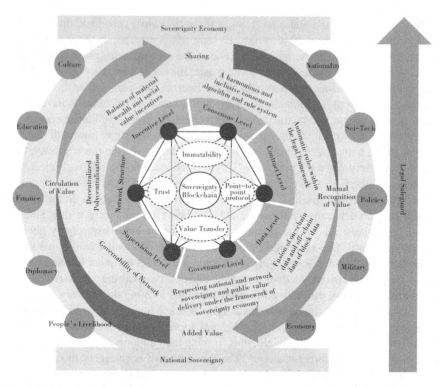

Fig. 1.4 Sovereignty blockchain

organic fusion of blockchain and block data is the most important breakthrough. It is not a simple technology fusion, but a people-centered approach to integrate humans and technology, technology and system, online and offline, and then integrate humans, technology, and society. This will be the beginning of the mutual authentication of human's first large-scale collaboration. Third, the sovereignty blockchain will elevate the Internet from a low level to a high one. Under the support and promotion of blockchain, the Internet will develop to be a "trilogy", namely, Information Internet, Internet of Value, and Orderly Internet. This significant elevation and evolution are determined by the features of blockchain technology. Although currently it still needs a period of time to explore, develop and improve the application of blockchain technology, the trend is unstoppable.

Blockchain and Sovereignty Blockchain. As sovereignty blockchain adds legal wings to the application of blockchain technology, blockchain develops from the rule of technology to the rule of system. The non-copiable data flow in the Internet is to be established in a framework that can be supervised and shared, so the institutional arrangement and governance system of the blockchain is stepped up. The sovereignty blockchain establishes rules through algorithms, and the parties can build mutual trust as long as the trust algorithm is used: there is no need to know others' credit, let alone the endorsement and guarantee of the third party, in order to establish a system that does not require trust accumulation, help establish a self-trust society and make data credit. By integrating the government into the network, sovereignty blockchain achieves the diversification of participants and gives play to the role of the visible hand of the government and the invisible hand of technology. White Paper, *Guiyang Blockchain Development, and Application* discriminate the sovereignty blockchain from eight aspects, including governance, supervision, network structure, consensus, contract, incentive, data, and application (Table 1.4). In terms of governance, sovereignty blockchain emphasizes mutual respect for cyber sovereignty among A Community with a Shared Future for Cyberspace, i.e., the delivery of public value under the framework of sovereignty economies, rather than the delivery of value over or without sovereignty. In terms of regulation, sovereignty blockchain highlights the regulated network and account, i.e., providing the technical control and intervention capability of regulatory nodes, rather than no regulation. In terms of network structure, sovereignty blockchain underscores the decentralized Polycentralization of the network, that is, providing the technical identity authentication and account management capability of each node under network sovereignty, rather than absolute "decentralization" or the formation of "super center". In terms of consensus, sovereignty blockchain emphasizes harmonious and comprehensive consensus algorithm and rule system, forms the wanted and required maximum

Table 1.4 Comparison between Sovereignty blockchain and other blockchains

Aspects	Sovereignty blockchain (Code + Law)	Blockchain (Code as Law)
Governance	A Community of a Shared Future for Cyberspace respects cyber sovereignty and national sovereignty and delivers public value within the framework of sovereignty economies	Without sovereignty or over sovereignty, the delivery of value that network communities agree on
Supervision	Supervision	No Supervision
Network structure	Decentralized Polycentralization	Decentralized
Consensus	A harmonious and inclusive consensus algorithm and rule system	Efficient consensus algorithm and rule system
Contract	Automatic rule generation mechanism under the legal framework of sovereignty economies, namely "code + law"	"Code as law"
Incentive	The balance between material wealth incentive and social value incentive	Material wealth incentive
Data	Based on the fusion of data on the chain and data off the chain	Data on the chain only
Application	Extensive application in various fields of economy and society	Limited to financial applications

Source Guiyang People's Government Information Office, *Guiyang Blockchain Development and Application*, Guizhou People's Publishing House, 2016

common divisor of each node, and provides the technical integration ability of various consensus algorithms, rather than the consensus algorithm and rule system that simply underlines efficiency first. In terms of contracts, sovereignty blockchain emphasizes the automatic rule generation mechanism under the legal framework of sovereign economies, namely "code + law", rather than just "code is law". In terms of incentives, sovereignty blockchain provides value weights and measures based on network sovereignty to achieve a balance between material wealth incentives and social value incentives, rather than simply material wealth incentives. In terms of data, sovereignty blockchain emphasizes parallel development of Internet of Things, big data, cloud computing, and other technologies to realize the fusion application of data on the chain and data off the chain, rather than only data on the chain. In terms of application, sovereignty blockchain emphasizes the extensive application in various fields of economic society, namely the integration and fusion of multi-field applications based on consensus mechanism, rather than only in the field of financial application. Sovereignty blockchain is an intelligent system based on the consensus, sharing, and co-governance of the Internet order. It is foreseeable that under the framework of sovereignty blockchain in the future, a new ecology will be formed in the Internet to change the established rules of the Internet world, and propose solutions to the global governance of the Internet, which will surely become an important turning point in the era of digitalization, networking, and intelligence.

1.6 From Unitary Dominance to Decentralized Co-governance

Humans are experiencing major historical transformations from the old era to the new era, not only in the technological revolution but in the organizational form and order system of the whole world. The world system of industrial civilization is characterized by human's division and cooperation in governance, while the new era will be characterized by division of rights and co-governance (Zhe 2018).

Evolution of Organizational Structure. Jeremy Rifkin, an American futurist and economist, said in *The Third Industrial Revolution*, "The world today is transforming from a concentrated second industrial revolution to a delayering third industrial revolution. Over the next half-century, the traditional, centralized operations of the first and second industrial revolutions will gradually be replaced by the decentralized operations of the third industrial revolution, with traditional, hierarchical economic and political power giving way to delayering power organized by social nodes". The organizational structure of human society has undergone long-term evolution and is becoming more diversified, including bureaucratic organization, delayering organization and network organization, from simple to complex, from vertical to horizontal, from closed to open, from tangible to intangible... When the organization is small and the value-creating activities are relatively simple, the power is concentrated in administrators at different levels, and the bureaucratic organization is an efficient organizational structure. With the expansion of scale, in order to address the change of value demand and the rigid structure of pyramidal organization, organizations came to develop toward horizontal operation order and to underscore the creativity of organizational value. Delayering organizations emerge when organizations move toward flexibility. Organizations become more open when they become delayering. The crisscross chain of value creation forms a network organizational structure. The organizational structure is more complex, the organizational form also more flexible, so the organization is more capable of coping with environmental changes and uncertainties and it is more open. The Internet-centered data revolution has deprived traditional public institutions of being the only ones to dominate data, but the real driving force for governance change.

Block Data Organizations are On the Rise. Sci-tech changes, especially the new generation of information technology, will promote the popularity and application of digital currency and digital identity. Whether it is a digital currency or digital identity, its popularity and application are bound to break through the "information island", and gather scattered point data and segmented bar data into a specific platform, and drive it to create a continuous aggregation effect. This aggregation effect reveals the essential laws of things via multidimensional data fusion, correlation analysis, and data mining, so as to make more comprehensive, faster, more accurate, and more effective research and prediction. This aggregation effect is called "block data". The continuous aggregation of block data forms block data organization, which will deconstruct and reconstruct the organization structure and trigger a new paradigm revolution. What is the paradigm revolution caused by block data organization and what kind of revolutionary change will it bring? In one word, it is a governance revolution that will change the future of the world. This is because the block data organization driven by digital currency and digital identity is essentially a decentralized and distributed organization mode under the control of impartial algorithms, which is called "decentralized and co-governed organization". The organization establishes a credible and tamper-resistant consensus and co-governance mechanism with three core technologies: Hyperledger, Smart Contract, and cross-chain technologies. Through programming and code, this mechanism solidifies the data flow formed by the superposition of time, space, and instant multidimensional, forming the technical constraint force that can be recorded, traced, confirmed, priced, traded, and supervised. With the development of blockchain, Smart Contract become extremely complex and autonomous. In essence, Dapp (decentralized application), DAO (decentralized autonomous organization), DAC (decentralized autonomous company), DAS (decentralized autonomous society), and other entities have become self-managed entities through increasingly complex and automated Smart Contract and connected to the blockchain through self-programming operations. When digital currency and digital identity meet the blockchain and make a perfect match, it marks that we have stepped into a new world. In this new world, the Internet is our computer. With the help of digital network and terminal, blockchain reconstructs the co-governance pattern of nation, government, market, and citizen, thus a multiparty co-governance model of global governance comes into being.

A New Paradigm for Sharing Organizations. Sharing is the biggest dividend brought by the Internet to mankind. What it brought is not only a new business model but also a revolution of the sharing society, stirring up the symbiotic and collaborative new paradigm of sharing organization. The relationship between people and organizations changes from an exchange relationship to a sharing one. Jeremy Rifkin believed, "There may not be simply value exchange in the future society, but value sharing. Everything in the past would have no value if not exchanged, but in the future, everything will have no value if not shared". Human evolution is inseparable from the law of symbiosis. Living beings are symbiotic; humans are symbiotic, so are organizations. Block data organizations break through the linear thinking mode of internal and external boundaries and traditional competition and turn from the logic of competition to the logic of symbiosis. The relationship between people and organization is a symbiotic relationship of value resonance, mutual dependence, co-creation and sharing, and the relationship between members featuring mutual subject, common resources, value co-creation, and profit-sharing. In addition, block data organizations achieve a collaborative transformation. The Internet and blockchain have brought about three basic changes to organizational management. The first is that efficiency no longer comes from the division of labor but from collaboration. The second is that the core of performance is to encourage innovation, not simply to examine performance. The third is the creation of a new organizational culture, with the interplay and mutual dependence of symbiotic members as its essence. The logic behind is what we call "collaborative efficiency". Block data organizations emphasize the altruism of data persons, which is beyond the hypothesis of economic man. Niccolò Machiavelli, an Italian political philosopher once said that it is not successful if it is not profitable for all involved, and it will not last long even if it is successful.

1.7   From Dependence on Objects to Dependence on Data

We cannot deny the existence of the data age, nor can we stop its advance, just as we cannot fight against the forces of nature. The data world is like a vast milky way. Humans are constantly exploring the data world, and the results of exploration drive the evolution of humans. From the agrarian age to the data age, data has undergone the evolution from digits to big data, from point data to block data, and from data to data rights. This is not only the evolution of data science dimension but the upgrade of human thinking paradigm. In the era of big data, individuals are both producers and consumers of data. When data production, data living, and data life become a reality and human intelligence and artificial intelligence are integrated, "natural persons" will evolve into "data persons". Data has covered and recorded a person's entire life from cradle to grave, and humans are dependent on data inveterately. The behavior pattern of humans will undergo a great change. When the dependence of people on people and on things[21] is not thoroughly eliminated, people begin to depend on "digits". The Orderly Internet and the sovereignty blockchain provide a new realistic possibility for freeing people from their dependence on "things" and "digits" in modern society, thus constructing and forming a new institutional model.

Dependence on People. In the *Economics Manuscript in 1857–1858*, Carl Marx divided the evolution of human society into three stages: the first stage is the pre-capitalist period, characterized by "the human productive capacity only developed in a narrow scope and isolated location", and the resulting social form is "the dependent society of humans". In such a society, the relationship between people "is either on the basis that the individual is not mature enough and is not yet free from natural blood ties to other people, or on the basis of direct rule and obedience". In this social form, the existence and development of humans are only in the space within the community, and humans cannot leave the community for a moment. In the era dominated by manual skills, that is, humans live by "human dependence", the most basic feature of human development is personal belonging on the basis of direct attachment to nature. This dependence suggests that human relations are local and singular, and therefore primitive or poor. "Neither the individual nor society can imagine free and full development, which is contrary to the original relationship between the individual and society (Marx and Engels 1979)". The development of human personality is still in its infancy.

Dependence on Objects. With the development of productive forces and social division of labor, the dependency relationship of people is replaced by the dependency relationship of things. "In this social pattern, a universal social material exchange, comprehensive relations, multifaceted needs and comprehensive system of capabilities are formed" (Marx and Engels 1979). That individuals get rid of the shackles of community, on the one hand, provides a possibility for the individual psychology to get rid of the low level and dependent harmony and move to the high level and autonomous harmony. On the other hand, individuals have to face a world full of uncertainty, process a variety of increasingly complex relationship and they are full of conflict and insecurity in their minds. Meanwhile, "dependence on objects", e.g., technologies, capital, and other things become the basic premise for people to be engaged in social production, which is embodied in the dependence of labor on capital and machinery. At this stage, technology, just like an engine, promotes the integration of people and the world. Technology is deeply embedded and reshaping

---

[21] The evolution of man is an important part of Marxist philosophy. In the Economic manuscripts of 1857–1858, Marx divided human evolution into the stage of human dependence, the stage of material dependence and the stage of human free and comprehensive development. "The dependence of man, which at first develops entirely naturally, is the original social form in which man's productive capacity develops only in a narrow range and in isolated locations. The independence of man, based on his dependence on things, is the second greatest form under which the universal system of social material exchange, comprehensive relations, multifaceted needs, and comprehensive capacities takes shape. The third stage is the free personality based on the full development of individuals and their common capacity for social production as their social wealth. The second stage sets the stage for the third." [Germany] Marx, [Germany] Engels. *Marx and Engels Collected Works (Vol. 46, Part 1)*, translated by Central Compilation and Translation Bureau, People's Publishing House, 1979, p.104.

human daily life practice and meaning generation, which has become a decisive force in human behavior, existence, creation, and social life. Human alienation is the inevitable development of human society. Every step forward is accompanied by a profound sense of alienation. Therefore, it can be said that man is an animal constantly alienated by virtue of technology.

Dependence on Data. Big data is a factor of production, an innovative resource, an organizational mode, and a type of right. The use of data has become a key way to increase wealth, and the assertion of data rights has become a significant symbol of digital civilization. In *Homo Deus: A Brief History of Tomorrow*, Yuval Noah Harari defines the new religion (Dataism) of the twenty-first century in this way, "The universe consists of streams of data, and the value of any phenomenon or entity lies in its contribution to data processing". Although the validity of the data-driven view is still open to question, we are indeed already in the ocean of big data and have formed an inextricable dependence on big data. Big data is intervening and changing the way of human production, life and existence at an unimaginable speed and depth. By emphasizing and advocating data productivity, big data establishes a new social relationship of combination, integration, and aggregation among data. Therefore, it is necessary to discard universally materialized dependency relationship, liberate people from attachment and subordination to objects, and make people the new people who rely on data for independent existence and free development. At this stage, human development is characterized by relatively personalized and free development on the basis of data dependence. According to the original meaning of bourgeoisie and proletariat, namely, the class with assets and the class without assets, those who hold the big data will become data bourgeoisie, while those who hand over the data will become data proletariat, thus causing the differentiation of human development. In the face of rising data tide, it is necessary to construct a rights protection system based on data rights, which is called "data rights system". The legal norm based on the data right system is called "data rights law". When data rights law and blockchain come together, the blockchain turns from the rule of technology to the rule of system. This kind of blockchain based on institutional arrangement and governance system is called "Sovereignty blockchain". It can be foreseen that sci-tech governance will bring humans to a new stage eventually. The key difference of this stage is that it has attacked the "human law" which has been unchanged for thousands of years and thus triggered an extensive and profound revolution in technologies, governance, and behaviors.

References

[US] Weibei E. Bijack. *Social and Historical Studies of Technology*. [US] Sheila Jasanov et al. *Handbook of Science and Technology*, translated by Sheng Xiaoming et al., Beijing Institute of Technology Press, 2004.

Tang Bin. The Internet is the Romance of a Group of People. *Chinese Business*, No. 5, 2015.

Chen Caihong. Ushering in the Fourth Industrial Revolution in Ignorance. *DuShu*, 2016, No. 11.

Ma Changshan. Disassembly and Reconstruction of the "Internet + Era" Rule of Law. *Exploration And Free Views*, 2016, No. 10.

Yu Chen. *Seeing the Future—People Changing the Internet World*. Zhejiang University Press, 2015.

China Mobile Research Institute (CMRI). *The 2030 + Vision and Demand Report*. Official website of the CMRI. 2019, https://cmri.chinamobile.com/news/5985.html.

Lin Dehong. Evolving Relationship Between Humans and Technology. *Science, Technology and Dialectics*, 2003, No. 6, p. 36

Lin Dehong. *Philosophy of science and technology*. Beijing university Press, 2004.

Liu Feng. *Internet Evolution*. Tsinghua University Press, 2012.

Yin Jianfeng. Digital Revolution, Data Assets and Data Capital. *China Business News Daily*, December 23, 2014, p.A9.

[US] Kevin Kelly. *The Inevitable: Understanding the 12 Technological Forces That Will Shape Our Future*, translated by Zhou Feng et al. Publishing House of Electronics Industry, 2016.

[US] Steven Kotler. *Tomorrowland*, translated by Song Lijue. China Machine Press, 2016.

[German] Marx, [German] Engels. *Marx and Engels Collected Works (Volume 1)*, translated by the Central Compilation and Translation Bureau of the Works of Marx, Engels, Lenin and Stalin. People's Publishing House, 2006.

[Austria] Victor Mayer-Schonberger, [Germany] Thomas Ramsey. *Das Digital*, translated by Li Xiaoxia, Zhou Tao. China CITIC Press, 2018.

[UK] Victor Mayer-Schonberger, Kenneth Cukier. *Big Data: A Revolution That Will Transform How We Live, Work, and Think*, translated by Sheng Yangyan and Zhou Tao. Zhejiang People's Publishing House, 2013.

[Japan] Mita Munesuke. Human and Society's Future, translated by Zhu Weijue. *Journal of Social Science*, 2007, No. 12.

Negroponte. *Being Digital*, translated by Hu Yong, Fan Haiyan. Hainan Press, 1996, p.15.

Wu Ning, Zhang Shujun. On Internet and Communism. *Journal of Changsha University of Science and Technology (Social Science)*, 2018, No. 2, p. 38.

Yao Qian, ed. *Blue Book of Blockchain: Annual Report on Dvelpment of China's Blockchain*, Social Sciences Academic Press, 2019.

Jiang Qiping. Digital Ownership Requires the Separation of Control and Use Rights. *China Internet Weekly*, 2012, No. 5.

[US] Jeremy Rifkin, *The Zero Marginal Cost Society*, translated by CCID Institute expert group, CITIC Press, 2017.

Li Sanhu. Data Socialism. *Journal of Shandong University of Science and Technology (Social Sciences)*, 2017, No. 6.

South Reviews Office. What Technology Wants. *South Reviews*, 2019, No. 26.

He Shen et al. Blockchain: The Future has Come. *People's Telecom Press*, November 15, 2019, p. 7.

Bai Shuying. On Virtual Order. *Study and Exploration*, 2009, No.4.

[US] Richard A. Spinello. *Century Ethics: Ethical Aspects of Information Technology*, translated by Liu Gang. Central Compilation and Translation Press, 1999.

[US] Jeff Stibel. *Breakpoint*, translated by Shi Rong. Renmin University of China Press, 2015, p. 20.

[US] Andreas Weigend. *Data for the People: How to Make Our Post-Privacy Economy Work for Us*. translated by Hu Xiaorui and Li Kaiping, CITIC Press, 2016, p.12.

[US] Jennifer Winter, [Japan] Ryota Ono. *The Future of the Internet*, translated by Zheng Changqing. Publishing House of Electronics Industry, 2018.

Wu Xiaobo. *Tencent Biography (1998-2016): Evolution of Internet Companies in China*. Zhejiang University Press, 2017.

Zhang Xiaomeng, Ye Shujian. *Blockchain: Principles, Construction, and Cases*. China Machine Press, 2018.

Duan Yongchao, Jiang Qiping. *The Origin of New Species: The Thinking Foundation of the Internet*. Commercial Press, 2012.

Pan Yunhe. The Human World, Cyber World and Physical World & the New Generation of AI. *Modern City*, 2018, No. 1.

[UK] George Zarkadakis. *In Our Own Image: Savior or Destroyer, The History and Future of Artificial Intelligence*, translated by Chen Chao. China CITIC Press, 2017.

Qiu Zeqi. *Toward the Digital Society, the Future is Coming: The Reconstruction and Innovation of "Internet +"*. Shanghai Far East Publishers, 2016.

He Zhe. Who Will Lead the World—Forms and Perspectives of Future Human Social Orders. *Journal of Gansu Administration Institute*, 2018, No. 4.

Chapter 2
Data Sovereignty Theory

With the fourth sci-tech revolution represented by digital technology and drastic changes in economic society, a large number of new human rights have emerged, among which "digital human rights" is the most prominent and important.

—Zhang Wenxian, Vice President and of Director of Academic Committee of China Law Society.

Digital sovereignty will become another room for leading powers to play games after border defense, coastal defense and air defense. The one who has the initiative and upper hand of data will win the future.

—Miao Wei, Minister of Industry and Information Technology of the People's Republic of China.

We must work together to ensure that big data, and the technologies that it enables, are harnessed for the benefit of humankind while minimizing the risks to development, peace and security and human rights.

—António Guterres, the Secretary-General of the UN.

2.1 Data Rights

With the evolution of data toward being resources, assets, and capital, the world will shift from the original system of property rights and creditors' rights to the system of data rights. The idea of data rights is a product of civilization progress as well as a new order for humans to step from industrial civilization to digital civilization. As a new right which is relatively independent of property rights, data rights together with human rights and property rights, constitute the three basic rights of human life in the future. Data rights, sharing rights, and data sovereignty form the core rights and interests of data rights, among which sharing rights is the essence. The idea of data rights is a significant force for order reconstruction and is of a special meaning for the common life of man.

© Zhejiang University Press 2021
L. Yuming, *Sovereignty Blockchain 1.0*,
https://doi.org/10.1007/978-981-16-0757-8_2

2.2 Human Rights, Property Rights, and Data Rights

In the era of big data, a new object that is different from and also beyond tradition has appeared in the vision of legal relations, which is "number". As humans step into digital civilization, the shift from "number" to "data" and then "data rights" is both the product of the times and the inevitable trend. Data rights is the basic human rights in the era of digital civilization, which means the basic rights of humans are protected, while data value is released. There are essential differences between data rights, human rights, and property rights. The differences in the subject, object, and contents of rights determine that data rights cannot be simply regulated according to property rights.

2.2.1 From Numbers to Data Rights

Numbers are not a natural existence, but the product of human social practice. In long-term practices, when humans compare the common features of different things in quantity, they understand the one-to-one corresponding logical relationship between abstract numbers and concrete entities—number in a creative way.[1] In ancient China, people paid special attention to "number". Taoism, represented by Lao Zi, explains the mode of the universe and the rule of changes of things with numbers. In ancient Greece, people's attention to "numbers" existed extensively in different schools, especially the Pythagoras School which showed an extraordinary worship of numbers. It raised numbers to the beginning of everything with an ontological significance and put forward the view that "number is the origin of everything", believing that number determines the form and contents of everything. The statement that "Number is the origin of everything" is not as physical as the opinions like atom or water is the origin of the world, but it means everything worldwide can be represented by numbers. This bold hypothesis of attempting to attribute the concept of all things to the human brain in the form of numbers and to understand the world with the ideas of numbers is exactly what big data technology comes to realize. Thus, numbers are not only the origin of the real world but the origin of the virtual world created by modern science and technology or the spiritual world of humans.

To some extent, "data is the extension and expansion of the concept of number, and the product of the progress of modern natural science, in particular, information science" (Tianping and Wenting 2016). The English word "data", first used in the thirteenth century, is derived from a Latin word "datum" which means a granted object. In modern society with the wide application of computers, digitization has been a reality, and the different forms of data, e.g., numbers, languages, characters, forms, and graphs have become the constituents of data. "Data is not limited to the

---

[1] Wang Yaode, Tan Changguo, He Yanzhen. Historical Phasing Observation of "Number" and Related Scientific and Technological Development. *Journal of Kaifeng Institute of Education*, 2017, No. 12, p. 11.

representation of specific attributes of things, but more importantly, it is the basis and foundation for deducing the motion and rules of changes for things" (Hong and Xinhe 2013). With emerging information technologies such as Internet of Things, cloud computing, and mobile Internet, more and more "things" are digitized and stored, thus a huge data formed. "Big data has once again unfettered human cognition of nature in the era of data. With massive data, humans capture every minute or second of changes in nature and record it in a more timely manner" (Tianping and Wenting 2016). As a new way to characterize the world, big data is profoundly transforming the forms of communication, organization, production, and lifestyle of human society, thus driving humans to the era of digital civilization.

In this era, humans begin to review their relationship with data and the rights of "data persons". Big data is a production factor, an innovative resource, an organization, and a type of rights. The use of data has been a major channel for wealth accumulation, and the idea of data rights has been a significant symbol of digital civilization. Data empowerment transfers the constitution of social forces from violence, wealth, and knowledge to data. There are many issues concerning rights and obligations in the whole life cycle governance of data, involving individual privacy, data property rights, and data sovereignty. Data rights, sharing rights, and data sovereignty have become new rights and interests in the big data era. Data rights are the greatest common divisor of sharing data to achieve its value. At present, there rose hot discussions on data rights and their ownership in the academic world, including personality rights, property rights, and privacy rights (Table 2.1). In addition, there are propositions like trade secret and intellectual property. However, traditional types of rights are insufficient to cover all the rights forms of data, and their ideas affect the integrity of data rights. While the digital era is multidimensional and dynamic, the design of data rights should not only reflect the unidirectional distribution of property rights of raw data but the dynamic structure and the rights of multiple subjects. Therefore, there comes a new type of rights—data rights, which cover all data forms, are actively playing a role and can be used by others.

2.2.2 Definition of Data Rights

The subject of data rights is a specific holder and the object is a specific data set (Table 2.2). In the specific legal relation of data rights, the holder refers to the specific holder. Data rights have different forms, such as data acquisition rights, data portability rights, data using rights, data revenue rights, and data modification rights. Therefore, it is necessary to determine the specific data rights holder according to the specific form and stipulated content of data rights. For the object of data rights, the single independent data has no value while only the data set which combines certain rules with independent value has its specific value, so the single data in the data set cannot be regarded as a separate data rights object. Therefore, the object of data rights is a specific data set.

Table 2.1 Data Rights Theories (Yu Zhigang. The Right Attribute of "Citizen's Personal Information" and the Thought of the Criminal Law Protection. *Zhejiang Social Sciences*, 2017, No.10, pp. 8–9; Ali Research. Wu Tao, Central University of Finance and Economics: Four Mainstream Views on "Data Rights and Ownership" in Legal Circles. Sohu.com. 2016. https://www.sohu.com/a/117048454_481893)

Doctrine	Proposition, reason, and defect
Personality rights	Proposition: Rights to personal data is not only personality rights but also specific and new personality rights
	Reason: First of all, in terms of the characteristics of the rights connotation, the rights to personal data protect personality interests, and data subjects have the rights attribute of controlling and dominating their own data, so they have specific connotations. Second, in terms of the richness of rights objects, the personal data of citizens include general personal data, private personal data, and sensitive personal data. Some of them, such as names, portraits, privacy, are specific personality rights and no longer protected by rights to personal data, while other data must be protected by the mechanism of personal data protection rights. Thirdly, in terms of the effectiveness of protection mechanism, if the rights to personal data is defined as property rights, there may be no need for protection. On the contrary, if it is regarded as personality rights, it ensures there will be no difference in calculation methods due to differences in personal status, which safeguards the personality equality on one hand, and citizens can claim compensation for mental damages under Article 22 of the *Tort Liability Law of People's Republic of China* on the other hand. Finally, from the perspective of comparative law, the personal data protection law worldwide mainly protects the personality interests of citizens
Personality rights	Defect: The personality rights of natural persons is exclusive and non-tradable. It cannot be treated as property even if it can generate economic value; otherwise, it will derogate the personality meaning of natural persons
Property rights	Proposition: Citizens' rights to the commercial value of their personal data is a new type of property rights, namely, "data asset rights"
	Defect: If it is simply regarded as a kind of property rights, the commercial value of the rights to personal data will be overemphasized and the protection of citizens' personal data is to be neglected. However, the latter one is the first goal of the legal system related to personal data and the most realistic need of citizens. In addition, if the "person" factor in "personal data" is ignored, it will inevitably be "talking about business in doing business", which impairs the equality of personality. "Because everyone has different financial situation and different values of information, but personality shall be protected equally and shall not be treated differently"
Privacy rights	Proposition: Citizens' rights to personal data shall belong to privacy rights, and citizens should seek relief through the relevant channels of privacy rights when their rights are violated
	Reason: First, the reason for protecting personal data is that infringement of citizens' personal data may violate their personal dignity and undermine the peace of their private life. However, if their personal data doesn't belong to privacy, even if others obtain it, nothing will affect the data subject and there will be no offence. Second, Article 2 of *Tort Liability Law of People's Republic of China* stipulates the protection of privacy rights, and through the extensive interpretation of privacy rights, it is sufficient to include all kinds of data protected by personal data into the concept of privacy, so there is no need to create independent rights to personal data

(continued)

Table 2.1 (continued)

Doctrine	Proposition, reason, and defect
Privacy rights	Defect: First, privacy rights highlight the protection of citizens' privacy and it focuses on passive defense, which is difficult to encompass the fact that a large number of citizens participate in various activities in modern society with the active use of personal data. Second, privacy rights are difficult to be compatible with the digital society. Privacy rights provide a kind of "absolute" protection of data. However, it is necessary and indispensable to collect, process, store, and utilize information in the digital society. Not only does the nation change from a simple protector to the largest user for data collection, processing, storage, and utilization but also third-party data practitioners such as companies and social organizations have gradually emerged. Third, the concept of privacy is vague. Making an issue of the subjectivity of "privacy" will make the judgment of privacy subjective, that is, the data subject determines whether it is the privacy or not. In fact, the key lies in "authorization", i.e., whether it has been consented to or not by data subjects

Source Rongyuan and Guanhua (2018)

Table 2.2 Features of data rights

Features	Outline
Subject of rights	The specific object to which the data points at and the one who takes charge of data collection, storage, transmission, and processing
Object of rights	The specific data set with regularity and value
Type of rights	A comprehensive right that integrates personality rights and property rights
Attribute of rights	Private rights attribute, nature of public power, and sovereignty attribute
Power and function of rights	A kind of sharing rights without excludability often manifested as "multi-ownership of data"

Data rights have private right attribute, nature of public power, and sovereignty attribute. Different from the traditional types of ownership, data rights reflect the diversity of ownership as a new type. Different types of data have different types of ownership, and data at different stages of the data life cycle also have different types of ownership. Data rights own private right attribute, nature of public power, and sovereignty attribute, meanwhile, including sovereignty that embodies national dignity, public power that reflects public interest, and data rights that highlight personal well-being. In the category of private rights, data rights are divided into personal data rights and enterprise data rights according to data subjects, and personal data resources and enterprise data resources are regarded as data rights objects. The nature of public power of data rights has its rich public and collective meanings. It is a kind of collective power with the state and government as the implementation subject, the maximization of public interest as the value orientation, and the strong maintenance of the public affairs participation order and it is self-expanding. The sovereignty attribute of data rights is reflected in the fact that data sovereignty is an essential part of national sovereignty. As a necessary supplement to national sovereignty, data sovereignty enriches and expands the connotation and extension of traditional national sovereignty, which is an inevitable choice for the nation to adapt to modern virtual space governance and safeguard the independence of its own sovereignty.

Data rights are a synthesis of personality rights and property rights. Data has both personality attribute and property attribute, but meanwhile, it is different from personality rights and property rights. The core value of data personality rights is to safeguard the human dignity of the data subject. In the era of big data, individuals leave "data footprints" behind in various data systems. Through correlation analysis, one's characteristics can be restored to form a "data persons". Recognizing the data personality rights is to emphasize that the data subject enjoys the rights that freedom is not to be deprived, reputation not to be insulted, privacy not to be pried, information not to be abused, etc. Meanwhile, the idea that "data is valuable" has become the consensus of the whole society, so it is necessary to endow and protect data property rights within the legal framework. As a new property object, data property should have the following five legal characteristics: certainty, controllability, independence, value, and scarcity.

2.2.3 Distinctions Between Human Rights, Property Rights, and Data Rights

Human rights are the only same symbol of humans and the greatest common divisor of all the people worldwide. Human rights mean "the rights that people enjoy and shall enjoy according to their natural attributes and social nature" (Buyun 2003). The person who enjoys human rights is not an economic person, a moral person, or a political person,[2] but a natural person who has biological characteristics and whose additional factors are all abstract. A person shall enjoy human rights just because he/she is a person. How did human rights come into being? This involves a very basic philosophical question of human rights. The theories on the origin of human rights are mainly customary rights theory, natural rights theory, legal human rights theory and utilitarian human rights theory, human nature origin theory, and moral rights theory.[3]

---

[2]The man that human rights refer to is not an economic man in the first place. The economic man aims at pursuing profits, and human rights will lack protection if everyone is an economic man. He is not a moral man secondly, because human rights have nothing to do with the existence and level of morality. In the end, he is not a political man. Although human rights are political, human rights will be restricted if they are used as a tool of political struggle.

[3]Customary rights theory is an empirical presumption of human rights represented by British Magna Carta, that is, the presumption of human rights "from customary rights to legal rights". Natural rights theory is a transcendental presumption of human rights carried forward by the French Declaration of Human Rights, that is, the presumption of human rights "from natural rights to legal rights", which is a classic theory on the origin of human rights. Legal human rights theory and utilitarian human rights theory believe that the formal or informal laws and regulations produce human rights,

Essentially, human rights are a kind of rights, "rights, human rights, legal rights and basic citizen rights are some inclusive concepts with generic relations" (Zhe 2004). Human rights have a relatively broad concept and connotation and protect a larger scope than legal rights or basic rights. With the in-depth development of economic society, human rights will continue to increase in dimension and category, and their connotation and extension will also extend.

The proposition of property rights is a new start of social civilization. Property rights are the domination of objects. Superficially, it is the domination of objects by people. In fact, it reflects the relationship between people. First, in essence, although property rights are the holder's rights to directly control "specific objects", it is not the relationship between human and objects but the legal relationship between people. Second, property rights are the property rights enjoyed by the holders toward "specific objects". Although it is a kind of property rights in nature, it is only the rights toward objects in property rights, different from rights toward people, creditor's rights. Third, property rights are mainly a kind of domination rights over tangible substance, that is, property rights holders can completely rely on their own will and exercise their rights without the intervention or assistance of others. The recognition of property rights is the recognition of the value created by individuals and individual autonomy rights after all. Therefore, property rights are human rights pertinent to objects. It is a kind of special and basic human rights. The recognition and protection of property rights mean people begin to take "people" as a new starting point and build a new coordinate of social civilization.

Different from the objects which are dominated exclusively in former civil laws, the domination of data is not exclusive objectively. This is determined by dematerialized data forms, and this feature is extremely similar to intellectual achievements. However, data is neither a property (movable property and immovable property) nor an intellectual achievement or rights. Data is a kind of object different from the "objects" with tangible forms, and the control over data is featured with non-exclusivity and non-loss (Aijun 2018). Because the ownership and control of the specific rights of property rights carried by the data are different from the possession and control modes of tangible substance, the property rights system suitable for tangible substance is not applicable to data. It can be considered that data rights do not belong to any traditional rights. Although some of its features are similar to other rights, it shall not be absorbed by expanding the property rights law or intellectual property law, and we should carry out special legislation on it according to consistent legislative habits (Table 2.3).

---

and the free and equal pursuit of personal happiness and welfare is the greatest value and goodness. Human nature origin theory holds that human nature includes natural attribute and social attribute, and the former one is the internal cause and basis of human rights, while the latter one is the external cause and condition of human rights. Moral rights theory holds that human rights belongs to the moral system and should be maintained by moral principles. Its legitimacy comes from human conscience.

Table 2.3 Distinction between human rights, property rights, and data rights

Project	Human rights	Property rights	Data rights
Subjects	Individual and collectivity	Specific persons	Specific rights holders include the specific object to which the data points at and the one who takes charge of data collection, storage, transmission, and processing (including natural persons, legal persons, unincorporated organizations)
Objects	Including the rights to objects, behaviors, spiritual products, information	Particular objects dominated by people; rights stipulated by the law	Data collection with certain rules or values; Exceptions may be provided by law
Contents	Personal and personality rights; political rights and liberty; economic, social and cultural rights; rights of vulnerable and special groups; international collective (or group) rights, etc	Ownership; Jus in re aliena (usufructuary rights and property rights for security)	Ownership; usufructuary rights of data; usufructuary rights of public data; sharing rights

Source Key Laboratory of Big Data Strategy (2018)

2.3 Private Rights, Public Power, and Sovereignty

2.3.1 Private Rights

The concept of rights is originated from the West, but "until the end of the Middle Ages, there was no word in any ancient or medieval language that could be accurately translated into what we call 'rights'" (Milne 1995). "In the Middle Ages, theologian Thomas Aquinas regarded 'jus' as a legitimate demand analytically for the first time, and named some legitimate demands of humans 'natural rights' from the perspective of the concept of natural law" (Wenxian 2011). "At the end of the Middle Ages, capitalist commodity economy made various interests independent and quantified, and the concept of rights came to be a general social consciousness". As a result, "jus" as "rights" is clearly distinguished from "jus" as "legitimate" and "jus" as "law" (Wenxian 2006). "The use of the Chinese characters "权利" as a term began in Japan". The word "权利" in Japanese was inherited from "jus" in Latin, "droit" in French, "recht" in German and "right" in English. It was first translated as "权理" and then as "权利" (Fan 2016). However, "asking a jurist 'what is rights' is as embarrassing as asking a logician a question known to all 'what is truth'" (Kant 1991). Although Kant's dilemma is a bit exaggerative, it can be used to illustrate the controversial and varied essence of rights, so Feinberg of the United States advocated that rights should be regarded as "a simple, undefinable and unanalyzable original concept" (Feinberg 1998).

At present, there are different theories on the nature of rights in academia. Among them, the representative ones that are closer to the essence of rights include qualification theory, freedom theory, will theory, interest theory, magic theory, and choice theory advocated by foreign scholars, as well as possibility theory and property theory held by domestic scholars. However, due to different or even conflicting historical contexts and positions, there has not been a unified paradigm of the theories on the essence of rights. Rights are characterized by their being historic, subject limiting, legitimate, material, reciprocal, and legal. In real life, rights are embodied or defined as legal rights, which means the legal system is of great importance to rights. From the perspective of legal system, rights are divided into fundamental law rights, public law rights, private law rights, social law rights, and mixed law rights. However, no matter what kind of rights it is, rights, mainly used to embody and safeguard personal[4] interests, are personal rights (Fan 2016). This kind of "person" is private fundamentally, i.e., the essence of rights is private.

Jurists in late ancient Rome defined private rights as the rights that involve personal interests, which are enjoyed by legal persons, unincorporated organizations, and natural persons (Xiuping and Jixiong 2013). There has not been a unified definition of private rights in China yet, and different scholars have different understandings and interpretations from different perspectives. Some think private rights refer to individual rights for meeting individual needs, while others think private rights are the rights in private law and they are the concept used as a term in the private law in civil society. Private rights are the rights confirmed by private law, pertinent to legal persons, unincorporated organizations, and natural persons and for accomplishing personal purposes. The subjects of private rights include citizens, legal persons, and other social organizations. Even if a nation does not exercise its authority or perform its duties in the name of public power, it still becomes the subject of private rights. Private rights are quite complicated. Some private rights have not or not yet been upgraded to legal rights, and they belong to a kind of personal behavior or personal freedom, while some private rights, confirmed by the Constitution or other laws, become legal rights. This divides private rights into two categories: non-legal private rights and legal private rights. The legal private rights are the main part and important contents of private rights, being explicitly protected by the Constitution and laws, Guangning (2010) and no one is allowed to violate in any form.

---

[4]The persons here are legally drafted individuals, that is, legal "persons", who include natural persons, legal persons and other unincorporated organizations.

2.3.2 Public Power

The issue of "power" has been perplexing philosophers and sociologists all over the world. Different scholars have different understandings of the concept of power. As Dennis Lang, a famous American sociologist said, "Power is an essentially controversial concept... People with different values and beliefs certainly have different opinions on its nature and definition" (Wrong 2001). British philosopher Robert Russell was the first to clearly define power. He believed that power is the ability of some people to produce expected or foreseeable effects on others. German sociologist Max Weber defined power as "the ability of someone or some to realize their will in social behavior, even ignoring the resistance of others who are involved in the behavior" in *Economy and Society*. American scholar Kurt W. Barker regarded power as a kind of "mandatory control under the conflicts of interests or values between two parties or all parties of individuals or groups" (Barker 1984). Such definitions are countless. Although they have their own points of view, these definitions fail to summarize the attributes and characteristics of power, especially state power (Daohui 2006).

The essence of power is public power. "The concepts of power and public power are mixed in many cases because power itself is public" (Aiguo 2011). "In a general sense, all power belongs to public power" (Xiaochun and Qiong 2012). On the one hand, from the perspective of the subject of power, the exerciser of power must be a public organ or a quasi-public organ (social organization). On the other hand, in light of the purpose of exercising power, power functions as the public interest protected by law. Therefore, to put it more accurately, power is public power. Public power refers to the power possessed by social communities (nation, social organizations, etc.) to manage public affairs, safeguard public interest of the state and society, and adjust the distribution of interests of all parties. "It aims at public interest and uses legal coercive power as a means. Public power is a necessary condition for the normal operation of a society and the basis of establishing and maintaining public order as well as ensuring social stability" (Guangning 2010). Public power includes state power and social power, which is usually divided into legislative power, judicial power, administrative power, and supervisory power.

"Public power, as the scepter of social life order, has long been regarded as the leader of social life" (Yanguo 2006). In the oriental society in particular, due to the influence of the concept "social standard", public power is regarded as the primary force which is dominant and decisive on the public. It is generally believed that the basis of public power is the public endowment of power and their recognition of the exercise of power. "State power is premised on people's alienation of rights and public recognition" (Zeyuan 2001). "In the ultimate sense, rights are the foundation of power" (Zeyuan 2001). Rousseau also pointed out in *Du Contrat Social* that state power is generated by every citizen transferring part of his/her private rights. It is also in this sense that "private rights are the origin of public power, and public power is the subsidiary of private rights" (Tao 2011). Without private rights, there is no need for public power to exist and develop itself. Meanwhile, public power and private rights are a pair of contradictions, which are both unified and oppositional, both interdependent and zero-sum. Their interaction forms the mutual connection between social interest groups.

2.3.3 View of Sovereignty

In ancient Greece and Rome, Plato and other sages realized the existence of sovereignty and studied the connotation of sovereignty. Although they did not explicitly put forward this key political concept of sovereignty, their studies of the emergence, function, and regime type centered on the state as well as governance of the state were essentially identical with the studies on sovereignty that we recognize today. They also laid a foundation for the explicit concept of sovereignty in the Enlightenment period (Plato 1999, Aristotle 1965, Cicero 1999). Aristotle, an ancient Greek philosopher, is considered the first sage to explain the idea of sovereignty. Although he did not explicitly use the concept in *Politics*, he already touched upon two major attributes of sovereignty, namely, external independence rights and internal supremacy power. The concept of sovereignty in the modern sense, originated in modern Europe, was the product of European economic and cultural development in the fifteenth and sixteenth centuries. French enlightenment thinker Bodin explicitly expressed sovereignty as "the absolute and permanent power of a nation to command" and "the highest power that is not subject to any legal restrictions on citizens and subjects" for the first time in *Les Six Livres de la République*. He believed that sovereignty is an indivisible, unified, permanent power of a nation which is beyond the law, and it is an absolute as well as eternal power.

After Bodin, Glausius, a famous Dutch international jurist and founder of modern international law, has improved the content of sovereignty on the basis of partial acceptance of Bodin's thoughts, revealing the duality of sovereignty, that is, the power with internal supremacy and external independence. Later, there came abundant connotations of sovereignty theories after the demonstration and development of modern political scientists such as Hobbes, Locke, Rousseau, Hegel, and Austin. However, sovereignty theory is a historical category with different connotations in different periods. Although many thinkers put forward their sovereignty theories and reached a certain consensus on the basic ideas of sovereignty, the definition of sovereignty has been inconclusive. *Oppenheim International Law* defines sovereignty as the supreme authority, which means "full independence, being independent no matter within or outside national territory" (Jennings 1995). China's international law circle generally adopts Zhou Gengsheng's definition of sovereignty, "Sovereignty is the highest power that a nation owns to deal with its internal and external affairs independently. Sovereignty has two basic attributes, which are supreme at home and independent abroad" (Gengsheng 2009).

Sovereignty can be regarded as national rights that embody the independence rights in the international community, and national power which embodies the supremacy power to manage domestic affairs. On the one hand, as national rights, sovereignty does not necessarily mean the mastery and possession of certain capability and power or the legitimate assertion of certain actions by other parties. Meanwhile, due to the influence and restriction of various subjective or objective factors, there are obvious differences between the freedom contained in sovereignty, and the scope and degree of this freedom are constantly changing. Especially with the development of international law norms and international legal order, the freedom contained in sovereignty is increasingly regulated and restricted (Zhou 2010). Therefore, in the modern international community, as national rights, sovereignty has a stable, clear, and fundamental connotation of independence and equal status. On the other hand, as a state power, sovereignty follows the basic law of power evolution. It is "jointly constructed by the people transferring part of their rights" and "is public in its source and representative in its exercise", Zhiying (2007) and it is a collection of national public power. In the big data era, sovereignty shows obvious collaboration and transferability.

2.4 Sharing Rights: Essence of Data Rights

Sharing, the effective use of data, represents data ownership in a comprehensive way. Different from property rights, data rights are no longer manifested as the rights of possession, but a kind of non-exclusive sharing rights often manifested as "multi-ownership of data". Once they rise from natural rights to a kind of public will, data rights will inevitably surpass their own form and be transferred to a kind of social rights. Sharing rights, the essence of data rights, are realized in public data rights and usufructuary data rights; thus, it is possible to separate data ownership and use rights. The proposition of sharing rights will become a new legal rule beyond the property law with the significance of digital civilization. It can be predicted that based on sharing, human civilization will move to a higher stage and enter an order constructed by sharing rights.

2.4.1 Sharing Rights and Possession Rights

Sharing and possession are the essential differences between data rights and property rights. Property rights include possession rights, use rights, earning rights, and disposition rights. Possession rights are the de facto control rights over property, and de facto control (possession rights) is the basis of ownership. Without possession rights, the exercise of the other three rights will be affected. Only by truly having possession rights, use rights, earning rights, and disposition rights can be better exercised. Both private ownership and individual ownership in human society take possession as their ultimate goal. However, with sharing economy, people realize that possession rights are not important, and what matters is whether they can be used by others. The essence of sharing is to share use rights and earning rights so as to obtain corresponding benefits (Zhe 2017). The transfer of use rights enables idle resources to be fully utilized, but its premise is that the right subject with possession rights has the will to transfer use rights, and its essence is "possession rights". Since the essence of property rights is possession, its root lies in the excludability of property, which determines that property cannot have multiple rights subjects. Meanwhile, possession becomes the only way to master property rights.

In the transfer of property rights, the existence of possession rights does not damage the interests of rights subject, who still enjoys control rights over the property. Different from property rights, data can have multiple rights subjects and can be copied indefinitely with extremely low cost and no loss. Under such circumstances, whether people have possession rights over data does not affect people's control and use of it. When they do not have possession rights, people can also exercise use rights, earning rights, and disposition rights of data. Once the use rights of data are transferred, the one that obtains the data will have the data completely, and the data will be out of the control of the initial holder. At this time, it is meaningless to enjoy the possession rights of data. In order to generate or maximize value, data must be shared with others, which will inevitably conflict with possession rights. Therefore, it is as important to emphasize the sharing rights of data rights as to emphasize the possession rights of property rights, which is the inevitable result of the development of "making the best use of data". In addition, the real value of data lies in unlimited reproduction at a low cost, which is the foundation for the development of digital civilization. It determines that sharing becomes the essential demand of big data era and that sharing rights becomes the essential rights of data rights. Breaking away from this foundation and applying the possession rights of property rights to data rights forcibly will hinder the application and development of data, violate or even destroy the original intention of data rights to protect, and develop data (Key Laboratory of Big Data Strategy 2018).

2.4.2 From "One Ownership for One Object" to "Multi-Ownership of Data"

"One ownership for one object" is the essential representation of the dominance of property rights. There is a greater variety of objects with the progress of science and technology. With the continuous increase of the types of property rights, the separation of property ownership and function has become increasingly complex, and the utilization of objects by people is also constantly changing. "One ownership for one object" is impacted by "multi-ownership for one object" and "multiple ownership "in reality. As the degree and form of human utilization of objects change constantly, the "multi-ownership for one object" and "multiple ownership" have also obtained some indirect acquiescence and vague permission of law in judicial practice, which breaks through the original meaning of "one ownership for one object". Different from the proposition of "one ownership for one object", the intangible attribute and replicability of data enable the existence of various interest forms. The creation of sharing rights enables data to have multiple subjects. Each subject does not share certain data rights, but has its independent and complete data rights, which forms a pattern of "not owning it but using it".

As data is replicable, non-consumable, and of special public, there can be "multi-ownership of data". This determines that giving any subject absolute domination rights over data will deviate from the development concept of sharing. With the progress of the times and sci-tech, when the cost of objects drops or even approaches zero, it will be no longer necessary to possess objects. This is especially true for sufficient data resources with zero marginal cost, so it will be an inevitable trend to advocate sharing of "multi-ownership of data". In the long run, resource scarcity in the traditional meaning will be addressed by sharing as scarce resources become rich. "When we think about it in terms of technology, there are few really scarce resources, and the real problem is how to use them" (Demandis and Kotler 2014).

2.4.3   Connotation of Sharing Rights

Sharing rights provide a data rights viewpoint with a balance between public welfare and self-interest for the era of digital civilization, conducive to stimulate people's creativity in digital civilization. The core of sharing rights is the balance of data interests. It violates the basic spirit of freedom and equality whether the public interests of data come before private interests or vice versa. Therefore, the imbalance in the distribution of data benefits will fundamentally discourage people's enthusiasm and initiative to create data wealth. The significance of creating sharing rights lies in that it revises the traditional rights and data views of "valuing self-interest and neglecting public interest", and advocates a data rights viewpoint that balances public interest and self-interest. Sharing rights are the premise of data use. This is not only the fundamental requirement for building digital civilization, but the essence of building a new social order as well (Rongyuan and Guanhua 2018).

Sharing rights, an integral part of the basic digital civilization system, take altruism as a fundamental basis. The distribution of public and private data interests is the core issue of digital civilization from the perspective of fairness. First of all, sharing rights must ensure the balance between public and private data interests in their distribution as well as fairness so as to streamline the relationship between public interests and self-interests of data subjects. Second, the distribution of public and private data interests is absolute, objective, and universal. No one can freely stipulate by his subjective will and any subjective, relative, or excessive interpretation is against fairness. Therefore, sharing rights are of great practical significance in a new order of digital civilization (Key Laboratory of Big Data Strategy 2018).

Sharing rights help coordinate the conflicts between different data subjects and provide the value-oriented basis for resolving the conflicts of data interests. Upholding the balance between public interests and self-interests of data, it provides a value-oriented basis for the construction of digital civilization system and makes fairness the primary value of its basic system. According to the principle of balanced distribution of public and private data interests, we establish legal norms to resolve conflicts between data subjects, improve the coordination mechanism of data interests among data subjects, unblock channels for data subjects to express their own demands for data interests, and tackle social crises caused by conflicts of data interests in order to enable data subjects to "do their best and get their places". Meanwhile, sharing rights help resolve the contradictions such as uneven distribution of resources, unequal opportunities, and social injustice brought about by data monopoly, address social unfairness and injustice, realize the optimal allocation and zero marginal cost of data resources, increase data wealth, and promote the coordinated development of economic society in the era of digital civilization.

2.5 Data Sovereignty and Digital Human Rights

Data sovereignty is a commanding height of data rights. As an increasingly important and urgent issue, data sovereignty has become the key concern of the nation, enterprises, and individuals in recent years. The core of data sovereignty is ownership identification. From the ownership of data, data sovereignty can be divided into personal data sovereignty, enterprise data sovereignty, and national data sovereignty. In practice, the relationship between data sovereignty and digital human rights has become a major theoretical and practical issue in the field of data sovereignty in the era of big data. Meanwhile, as data sovereignty has become an area of overlapping interests, there are challenges such as unclear data ownership, confusion in data circulation and utilization, and personal information leakage. In such a context, blockchain provides a feasible technical solution to data sovereignty protection. With its inherent characteristics of decentralization, tamper-resistance, traceability, and high reliability, blockchain resolves data sovereignty definition and ownership, thus breaking the monopoly of data sovereignty and benefiting humans with its maximum functional value.

2.6 Differences Between Sovereignty and Human Rights

The relation between sovereignty and human rights does not only involve theoretic solutions to relevant problems but depends on a consensus to a certain extent in practice. Tension between sovereignty and human rights had already existed when their concepts came into being. The tension began to expand with a progressing society, causing further opposition and conflicts between sovereignty and human rights. However, the tension between sovereignty and human rights needs to be dispelled and the opposition needs to be unified with human pursuit of a dignified life. Therefore, sovereignty and human rights should eventually be unified in theory and in practice.

2.6.1 Tension and Conflict Between Sovereignty and Human Rights

The relation between sovereignty and human rights is the focus of modern international relations and international politics, and it is also a major issue that has sparked heated arguments in the field of international human rights since the end of the Cold War. "The relation between the two is not only related to the rights and interests of every unit person on the earth, but involves multi-level interest distribution points such as domestic politics and international issues" (Huan 2015). In the current international political landscape, the relation between sovereignty and human rights presents a crisscrossing "cluster" state, with conflicts between domestic sovereignty and domestic human rights, as well as conflicts between international human rights and external sovereignty. Under the guise of "human rights are more important than sovereignty", the US-led Western developed countries have either wantonly propagated their values or groundlessly accused some developing countries of violating human rights, and have used this as an excuse to launch human rights diplomacy or humanitarian intervention toward these nations. Meanwhile, many developing countries have also expressed their willingness to settle differences with western nations on human rights through dialogues. In this way, two tit-for-tat views "human rights are more important than sovereignty" and "sovereignty is more important than human rights" were born. The reason for the opposing views is that there is tension between sovereignty and human rights, including internal tension and external tension.

As far as internal tension is concerned, first, in view of the subject of sovereignty and human rights, sovereignty is often enjoyed by the ruling clique which acts on behalf of the nation, while the holders of human rights are ordinary people who account for the majority of society in most cases, including various vulnerable groups. Because the ruling clique has its own thinking pattern, value proposition, operation logic, and interest orientation, its attitude is often different from that of ordinary people. As a result, the relation between human rights and sovereignty is always strained due to the different status of their subjects. Second, in terms of the contents of sovereignty and human rights, sovereignty includes rights and power, while human rights only include rights. It is the above-mentioned different contents that determine the value orientations of sovereignty and human rights. There must be internal tension between control and anti-control due to different value orientations.

As far as external tension is concerned, from the domestic level, whether the government allows citizens to have resistance rights or disobedience rights and its attitude toward these two rights are the main manifestations of the tension between sovereignty and human rights. In the anti-Vietnam War movement in the US in the 1960s, the dilemma between citizens' resistance rights and maintenance of legal order was fully reflected. In the international landscape, the gap between different countries in the nature and priority of sovereignty and human rights is a concrete manifestation of the external tension between sovereignty and human rights. In terms of the nature and contents of sovereignty and human rights, Western developed countries believe the philosophical basis of human rights is "natural rights". In contrast, in the political concepts of many countries, basic human rights and freedom are often regarded as being granted to the people by the nation, and the nation also legally determines the degree of freedom and rights enjoyed by the people.

2.6.2 Communication and Agreement Between Sovereignty and Human Rights

Since there is huge internal and external tension between sovereignty and human rights, only by seeking communication between the two and resolving the tension can we ensure the sustained, effective, and harmonious development of sovereignty and human rights. The harmonious coexistence of sovereignty and human rights is not only the ideal goal that humans pursue unremittingly but a key benchmark for system construction and perfection. We should explore the common philosophical basis between the two for the harmonious coexistence of sovereignty and human rights. Comprehensive knowledge and understanding of the philosophical basis of sovereignty and human rights need investigation from different dimensions, including cultural status, historical development, and economic and social development of different nations. Through extensive and comprehensive investigation, we can find that "humanism" is the philosophical basis for the coexistence of sovereignty and human rights. Starting from this philosophical concept, it can be easily seen that both sovereignty and human rights serve freedom, happiness, and interests of humans, and they are not an end in themselves. In addition, compared with ultimate human goals, i.e., freedom, happiness, and interests, sovereignty and human rights are equal and homogeneous in origin.

Although there is a common philosophical basis of "humanism" and homogeneous views between sovereignty and human rights, due to practical interests and different historical understandings, there has always been tension between sovereignty and human rights, which requires the representatives of sovereignty and human rights to discuss and communicate on the basis of equality. Habermas, a prestigious German philosopher and sociologist offered a realistic and feasible way in his communicative action theory to resolve the tension between the two. In an international community with numerous sovereignties, sovereignty barriers and human rights tools will be inevitable if there is no universally applicable set of sovereignty and human rights norms. Habermas believed "One way out of this dilemma is to regulate the strategic interaction normatively, and the actors should reach an understanding" (Habermas 2003). Norms that are binding on each member cannot form via a single subject. Only by considering and coordinating the interests of all relevant subjects and a consensus among subjects through rational examination and public debate can norms with universal binding force be developed. The resolution of the tension generally depends on the formation of sovereignty and human rights norms, and it is necessary to ensure "everyone has equal opportunities to exercise the freedom of communication to express his attitude toward critical and effective claims" (Habermas 2003).

In the practice of sovereignty and human rights, the external tension which came to be formed in international practice between the two especially needs to be weakened and dispelled by rational consensus through intersubjective negotiation and communication on the premise of mutual respect and equality, that is, sovereignty cooperation replacing sovereignty barrier and human rights dialogue replacing human rights confrontation (Lin 2013). On the one hand, developed countries should fully respect and consider the serious concerns of the developing countries and stop using human rights as a weapon to promote power politics and hegemonism. On the other hand, developing countries should ensure the independence of national sovereignty, territorial integrity, inviolability of national dignity, and solid political power, as well as be aware of the steady "expansion" of the interests of developed countries while improving the political and legal system guarantee for long-term stability and sustainable, healthy economic and social development, as well as the development of human rights. The relation between sovereignty and human rights is a dynamic balance with mutual promotion, interdependence, and unity of opposites. The priority of the two should be weighed based on different national situations, taking into account various factors such as political interests, economic and social development, ethnic and cultural traditions of different nations to formulate a better balance point.

2.7 Human Rights of the Fourth Generation

According to the wide-accepted statement in academia, the pattern of human rights worldwide has achieved three historic leaps so far in three successive generations of human rights (the first generation, the second generation, and the third generation of human rights). At present, the fourth generation, a new generation of human rights led by "digital human rights", is approaching. The relation between digital human rights and the first three generations is not of totally replacing, let alone opposing. They are progressive and expansive, transforming and upgrading. The four generations of human rights constitute the human rights system of the new era (Table 2.4).

2.7.1 The First Generation of Human Rights

The first generation of human rights formed during the French Revolution in 1789. With freedom as its main part, individual autonomy as its fundamental purpose, and classical liberalism as its ideological basis, it is also commonly called "the right to freedom" (Guanghui 2015). The concept of the first generation of human rights tends to laissez-faire economic and sociological theories. Its liberal philosophy which is compatible with individualism is the inheritance of Britain, America, and France in the history of human rights and proposition of developing revolution (Guanghui 2015). As for schools of thought, the views of the first generation of human rights

Table 2.4 Comparison of four generations of human rights

Program	First generation of human rights	Second generation of human rights	Third generation of human rights	Fourth generation of human rights (Digital Human Rights)
Generation background	Born in 1789 during the French Revolution, the historical background was the bourgeois revolution against feudalism and autocracy	Born at the beginning of the twentieth century after October Revolution in Russia, the historical background was the socialist revolution that opposed capital exploitation and eliminated the polarization between the rich and the poor	Born in the 1950s and 1960s during the liberation movement of colonies and oppressed people, the historical background was a national revolution for national independence, national liberation, and political democracy	The fourth industrial revolution represented by digital technology is born with rapid changes in economy and society, and the historical background is an information revolution
Human rights propositions	Advocating rights to life, to personal liberty, freedom of belief, of religion, of speech and publication, of assembly and association, of movement and residence, freedom from arbitrary detention, freedom of undisturbed correspondence, and political rights such as the right to vote, with special emphasis on the inviolability of property rights	Advocating rights to work and subsistence, besides retaining the contents of the first generation of human rights, also putting forward the rights to work, rest, medical and health care, education, maintenance of a moderate standard of living, and the rights of workers for unity	Advocating rights to peace, development, to the environment, to national self-determination, and to the common heritage of humans	Advocating the autonomy of data information, right to acquire, to express, and to use data information fairly, right of data information privacy and data information asset rights, etc
Core	To safeguard individual freedom in legal form, oppose the government's improper interference in individual freedom and rights with political power, and require the nation to burden the obligation of negative omission	To require the nation to provide basic social and economic conditions to realize individual freedom and emphasize national obligation to take positive actions toward human rights	Because of its joint nature, it can be called "related right" or "joint right". It has a collective nature and focuses on national determination and development	To eliminate human rights threats such as algorithm discrimination, data divide, social monitoring, algorithm hegemony, etc., and to enhance human autonomy in the digital age, as well as protect human rights of "digital man"

Source Guanghui (2015), Yanping (2015) , Changshan (2019)

include mainly the view of Humans Rights for Purposes,[5] of Natural Human Rights,[6] of Human Rights of Wills,[7] of Natural Human Rights,[8] of Religious Human Rights,[9] of Utilitarian Human Rights.[10] From the perspective of social basis, the first generation of human rights is formed on a combination of many factors. It is a human rights theory created in the competition with autocratic nations. It aims at protecting individual freedom and rights, opposing improper national interference through political power, and requiring the nation to undertake the obligation of negative omission. Therefore, it is called "negative human rights". In other words, the establishment of human rights is just to support the basic operation of society, with the means of restricting the procedure and scope of government power operation by affirming individual rights. The primary mission of a nation is to create a relaxed environment for free competition and to maintain social order, not to interfere too much in social production and economic life. Thomas Paine, a famous American thinker, said, "The government that manages the least is the government that manages the best. The government itself does not have any rights, but is only responsible for obligations". The first generation of human rights advocates those which include the right to life, to personal liberty; freedom of belief, of religion, of speech and publication, of assembly and association, of undisturbed correspondence; freedom from arbitrary, illegal arrest and detention, with special emphasis on the sanctity of property rights (Guanghui 2015). Its focus is to safeguard individual freedom in the legal form, which embodies the personal liberalism prevailing in the seventeenth and eighteenth centuries, and lays a solid foundation for political rights and citizen's rights (Yanping 2015). However, with historic changes, especially the strong influence of capitalism on individuals and social fields, people's expectations of the government and the function of the government have changed greatly, thus adding new connotation and significance to the concept of human rights in modern times (Guanghui 2015).

---

[5] View of Human Rights for Purposes holds that humans themselves are ends and all people have their rights, represented by Kant and some human rights theorists of the Neo-Kantian School.

[6] View of Natural Humans Rights is a kind of human rights theory with continuous impact advocated by many thinkers in the Eastern and Western societies since ancient times. It has been confirmed in documents like American Declaration of Independence and French Declaration of Human Rights.

[7] View of Human Rights of Wills, also called the internal drive theory of human rights, advocates humans have internal values, that is, the dignity of personality comes from the free wills and rationality of humans, represented by Hegel and Fichte, etc.

[8] View of Natural Human Rights, also called the theory of natural rights or the theory of human rights instinct, prescribes human rights are the natural rights of humans and the self-evident rights that humans have due to human instinct in a natural state, represented by Aristotle and Cicero, etc.

[9] View of Religious Human Rights, developed from the human rights theory of Christianity, Catholicism and other major religions since Roman times, holds that humans have rights because they are sons of God, represented by Marcus Aurelius Carus and Joseph Alois Ratzinger, etc.

[10] View of Utilitarian Human Rights, also called the interest-driven theory of human rights, holds that rights are generated from interests, and the birth of human rights is driven by interests. This theory has been systematically expounded in the theory of Jeremy Bentham, founder of utilitarianism theory. However, the View of Utilitarian Human Rights is opposed by the supporters of View of Human Rights of Wills, for instance, Hegel did not agree to understand the nature of rights from the perspective of human interests.

2.7.2 The Second Generation of Human Rights

Originated after the October Revolution in Russia,[11] the second generation of human rights was commonly called "social rights". The term refers to active national intervention in the whole economic society to protect the rights of social or economic life of all the people (Chongde 2009). As the second generation of human rights, social rights are originated from socialist criticism and amendment of capitalism. After the first generation of human rights progressed for over 100 years, in the second half of nineteenth century, especially at the turn of the nineteenth and twentieth centuries, capitalism swept across the world with the power of the Industrial Revolution and changed entire human civilization and the way of life. With highly developing capitalism and the steady development of monopoly enterprises, maladies of capitalism, such as unemployment, poverty, inflation, and food crisis, have cast a huge shadow on society. The legal pillar of capitalist society and the foundation of free human rights, property rights, and freedom of contract were overwhelmingly conducive to property owners and utterly unfavorable to proletarians (Guanghui 2015). In this way, all freedom and rights were likely to become the imagined satisfaction without any practical significance. However, unemployment and poverty were not caused by personal laziness, but the inevitable result of the economic structure of capitalist society. Unemployment and poverty should be solved by society and even the state (Akira 2001). As a result, a socialist trend of thought came into being to address the maladies of capitalism and transform the current unfair and unjust society. The thought claims to require the state to protect and improve the lives of laborers and interfere in the exploitation of capitalists, in order to ensure that laborers participate in the production and distribution of values fairly (Guanghui 2015). The second generation requires the state to provide basic social and economic conditions for individual freedom, emphasizing the national obligation to take positive actions toward human rights, which is called "positive human rights" (Yanping 2015). It is a theoretical proposition centered on human rights such as the right to subsistence and work, characterized by a shift from the pursuit of individual rights to the demand for collective and hierarchical rights. It focuses more on economic, social, and cultural rights. In addition to retaining the contents of the first generation, the second generation also proposes the rights to work, rest, medical and health care, education, maintenance of a moderate standard of living, and solidarity of workers (Guanghui 2015).

---

[11] The October Revolution in Russia is a great socialist revolution by the Russian working class in conjunction with the poor peasants under the leadership of the Bolshevik Party. The triumph of the October Revolution ushered in a new era in human history and paved the way for the victory of proletarian revolutions, colonial and semi-colonial national liberation movements worldwide.

2.7.3 The Third Generation of Human Rights

The third generation of human rights, born during the liberation movements of the colonies and oppressed peoples in the 1950s and 1960s, is commonly called "social joint rights", focusing on the self-determination and development of nations, reflecting the demands of the third world nations to redistribute global resources and their choices in facing major threats to human survival after World War II (Yanping 2015). It discusses the collective "joint rights" related to human living conditions, with its main contents of the right to peace, development, environment, national self-determination, and common heritage of humans, etc. As these rights, to be realized, can only rely on the joint efforts of all participants (including individuals, nations, public and private institutions, the international community), they are also regarded as "collective human rights".[12] There are great differences between the third generation and the previous two generations in terms of subject scope. If the first two generations are based on claims arising from the relationship between individuals and their nations and between groups and their nations within the national boundary, then the third generation has changed its direction of rights, which means rights are no longer the demands made by individuals to nations, but by one nationality to another nationality, one nation to another nation, or even one nation to all other nations or the international community (Guanghui 2015). The essence of the third generation lies in its joint nature, which can be called "related right" or "joint right". As a collective nature, it surpasses the previous concept of "individual human rights" and is recognized as collective or even social justice. Its views mainly include the Absolute View of Human Rights,[13] Relative View of Human Rights,[14] Confucian View of Human Rights,[15] Liberal View of Human Rights,[16] Collective View of Human Rights,[17] Asian Value View of Human Rights,[18] and a Critical Multicultural View of Human

---

[12] Wang Guanghui. *Comparative Constitution*. Wuhan University Press, 2010, p. 89.

[13] The Absolute View of Human Rights advocates that human rights is heaven-born, natural, inalienable, unconditional, and unchangeable. Representative figures include Blake, Douglas, Mackler Joan, Rostow, Blake, etc.

[14] The Relative View of Human Rights advocates that human rights is social, moral, transferable, conditional, and variable. Representatives include Brandeis, Dewey, Hooker, Bodenheimer, Schwartz, etc.

[15] Confucian View of Human Rights is a human rights theory established on traditional Confucian philosophy and morality. Its representative figures include Prof. Chung-Ying Cheng and Du Gangjian.

[16] The Liberal View of Human Rights, based on individual rights, is represented by John Bordley Rawls, Ronald Myles Dworkin, and Nosik, etc.

[17] Collective View of Human Rights, opposite to the liberal view of human rights, pays attention to the collectivism of human rights and believes that the collectivism of human rights is more important than the individuality. Its representatives include McIntyre, Walsh, Aitcioni, Selnik, Granton, Bailey, etc.

[18] Asian Value View of Human Rights, highlighting the significance of the human rights value of traditional Asian culture to the development of human rights, is established on the basis of reflecting on the monopoly of Western culture in the discourse power of human rights and trying to find explanations in traditional Asian culture. Its representatives include Lee Kuan Yew of Singapore,

Rights.[19] In addition, the third generation also stresses there may be differences in the connotation of human rights in different traditional cultures. For example, the "Asian value human rights view" held by Lee Kuan Yew and Mahathir in the 1980s is not the view of Western human rights. The third generation emphasizes the focus of the concept of human rights also differs under different social and economic conditions. Third world nations generally believe that development is the foundation of all kinds of human rights, and there would be no human rights without development. Although the third world nations do not deny the equal importance of civil and political rights as well as economic and social rights at the theoretical level, in practice, limited by limited resources, backward medical care, inadequate education, and past colonial exploitation, these nations can only give priority to economic and social rights over civil and political rights. Otherwise, for people who are unable to maintain basic living conditions, it will be redundant no matter how complete the civil and political rights are Guanghui (2015).

2.7.4 Digital Human Rights

With the fourth industrial revolution represented by digital technology and rapid changing economic society, human rights are undergoing profound digital reshaping, thus breaking the existing development pattern of "three generations" of human rights, and creating digital human rights, the fourth generation of human rights. Digital human rights "takes the relation between production and life in dual space as the social foundation, the digital information orientation and related rights and interests of humans as the expression form, and all-round development of humans in a smart society as the key demand" (Changshan 2019). It aims at eliminating human rights threats such as algorithmic discrimination, social monitoring, digital divide, and algorithmic hegemony, improving people's autonomy in the digital age, and deepening the human rights protection of "digital man". Digital human rights have abundant connotations, "including not only 'realizing human rights through digital technology', 'human rights in digital life or digital space', but also 'human rights standards of digital technology' and 'legal basis of digital human rights'" (Wenxian 2019).

Digital human rights, originated from information revolution, underwent fundamental changes in connotation and logic. There is a revolution behind each generation from the first to the third. Behind the first generation of human rights is the bourgeois revolution against feudalism and autocracy, behind the second generation is a

---

Mahathir Mohamad of Malaysia, Choi Chung-ku of South Korea, Kochu Nobuo, Yusada Nobuyuki, and Keio Suzuki of Japan.

[19]Critical Multicultural View of Human Rights, forming on the basis of "multiculturalism" that appeared in the 1970s, pays attention to the utilization of cultural resources in human rights research, the value premise of human rights, the substantive basis, and procedural basis of human rights. Its representatives include Alain Supiot, Jürgen Habermas, Onuma Yasuaki, and Yasuhisa Ichihara, etc.

socialist revolution against capital exploitation and the elimination of polarization between the rich and the poor, and behind the third generation is national revolution for national independence, national liberation, and political democracy. Behind digital human rights today is an information revolution, bringing about ideological emancipation and system innovation to humans. However, information revolution subverts the production and life in a traditional industrial and commercial era in the form of technological revolution, rather than armed struggle. Besides, the internal logic of digital human rights is different from that of the previous three generations. In terms of economic security, survival, or political participation, the first three generations share two common features: one is to express demands based on human biological attributes, and the other is to unfold within the logical framework of physical space. However, the demand for changes and objective development of digital human rights is neither an expansion of human rights in traditional industrial and commercial era nor an increase in the number and types of rights, but a fundamental turn of human rights in the digital era.

Digital human rights upgraded the quality of human rights. Human rights are the rights humans shall enjoy according to their nature, and everyone should be treated in accordance with human rights. This is both the key to the morality and universality of human rights, as well as the core value of human rights. The development and reform of human rights at all stages lead to the upgrading and transcendence of existing core values of human rights. The second generation of human rights surpassed the first generation and moved toward a more substantial view of social, cultural, and economic rights, while the third generation surpassed the second generation and moved toward a collective view of rights that focuses on survival and development. The same is true of current "digital human rights" (Changshan 2019). Compared with traditional human rights, digital human rights are not an expansion of traditional human rights, but the upgrade of human rights quality brought about by a smart society and information revolution. It faces a tech-revolution that breeds both opportunities and challenges. It is necessary to curb the negative risks of "digitalization, networking and intelligence", to transform its achievements into the free development ability of humans, break through the biological boundaries of humans as well as the physical time and space stipulated by God for humans, and thus approach the value and dignity of humans.

2.8 Blockchain Reshapes Data Sovereignty

Data sovereignty includes personal data sovereignty, enterprise data sovereignty, and national data sovereignty. In practice, there are many challenges such as ambiguous definition and ownership of data sovereignty. Against this backdrop, blockchain technology provides a feasible technical solution to data sovereignty protection. It is expected to address data sovereignty with its characteristics of decentralization, tamper-resistance, traceability, and high reliability.

2.8.1 Re-Understanding Data Sovereignty

As data has become a national fundamental strategic resource, any illegal intervention in data by any subject may lead to the infringement on a nation's core interests. Based on the needs of personal privacy, industrial development, national security, and government law enforcement, data sovereignty emerges (Bo 2017). Data sovereignty "involves the generation, collection, storage, analysis and application of data, which directly reflects the value of data in information economy" (Wei and Yue 2015). At present, the theoretical circle has not reached a consensus on the concept of data sovereignty, but in general, there are two understandings of data sovereignty in a broad sense and narrow sense. According to Joel Tsuen Ziman, a professor at Tufts University, America, data sovereignty in a broad sense includes national data sovereignty and personal data sovereignty, while data sovereignty in a narrow sense only refers to national data sovereignty. For research and understanding the significance of the relationship between data and individuals, enterprises, and nations, we adopt the broad concept and re-divide data sovereignty into personal data sovereignty, enterprise data sovereignty, and national data sovereignty according to the attribution of data.

Personal Data Sovereignty. Personal data sovereignty, also named rights to personal data, refers to the rights of the data subject to dominate and control its personal data according to law and the rights to exclude infringement by others.[20] Personal data sovereignty includes not only the rights of personal privacy data not to be violated by others but also the rights of personal property rights and personality rights not to be violated, as well as the rights of personal free will not to be fettered. According to the legal definition of personal data and the particularity of its right object, compared with intellectual property rights and general property rights, personal data rights are featured with universality, separation, invisibility, duality, and expansibility. Personal data sovereignty is an independent new right in big data age. Different from traditional personality rights, property rights, and privacy rights in terms of object, content, and form, it plays an irreplaceable role.[21] From the contents of rights, personal data sovereignty generally includes the right to correct data, to delete data, to seal up data, to know data, to keep data confidential, and to claim data compensation.[22]

Enterprise Data Sovereignty. Enterprise data, actually controlled and used by enterprises, is expressed in the form of symbols or codes. It includes not only financial data, operation data, and human resources data which reflect the basic situation of the enterprises but also user data legally collected and utilized by enterprises (Dan

---

[20]Qi Aimin. *Research on Principles of Personal Data Protection Law and Legal Issues of Transnational Circulation*. Wuhan University Press, 2004, p. 110; Liu Pinxin. *Internet Law*. China Renmin University Press, 2009, p. 93.

[21]Qi Aimin. *Research on Principles of Personal Data Protection Law and Legal Issues of Transnational Circulation*. Wuhan University Press, 2004, p.109.

[22]Wang Xiuxiu. *Personal Data Right: The legal protection mode from the perspective of social interests*. Doctoral Dissertation of East China University of Political Science and Law, 2016, p. 61.

2019). As there is an increasing call in all sectors of the community for the protection of enterprise data and its related rights and interests, the concept of enterprise data sovereignty has been put forward. Also named rights to corporate data, enterprise data sovereignty refers to the right of an enterprise to possess, use, interpret, manage, and protect kinds of valuable data resources generated in its operation and management, and not to be violated by any organization, unit, or individual. From the perspective of legal attributes, enterprise data sovereignty is neither property rights nor simply a kind of intellectual property or property rights but a bundle of rights formed by the collection of different rights and interests. At present, there are mainly two types of claims in China, "one claims that enterprises have extensive rights to the user data they collect; the other is to classify the data held by enterprises and then to claim that enterprises have rights to some types of data" (Wei 2019).

National Data Sovereignty. With the rapid development and wide application of the new information technology, data space has become the fifth space for human existence. Cross-border data flow and storage come to be routine and convenient, posing a serious threat to national data security and being incorporated into a nation's core power. National data sovereignty refers to the supreme power of a nation to independently possess, process, and manage its own data and exclude the intervention of other nations and organizations. It is "an inevitable requirement for nations to safeguard national sovereignty and independence, combat data monopoly and hegemonism in the era of big data" (Gang et al. 2017). National data sovereignty, with the attributes of times, relativity, cooperation, and equality, mainly includes data jurisdiction, data independence, right to self-defense over data, and right to equality over data. As a product of national sovereignty in the era of big data, national data sovereignty emerges on the existence of data space and embodies and extends national sovereignty in data space. In the data space with vague borders, the claim of national data sovereignty is not only an effective restriction on the power abuse of big powers but a major manifestation of international security interests and a reflection of the concept of peaceful coexistence.

2.8.2 Sovereignty Under Blockchain

Blockchain shifts human society from the Internet of Information era to the era of Internet of Value. While bringing about an efficient system of credit creation and value circulation, the blockchain has also raised the issue of sovereignty to a significant position. At present, the issue of sovereignty under the blockchain not only becomes a new arena for competition among individuals, enterprises, and nations but also gives birth to new bottlenecks and risks of global economic growth and social development. Sovereignty under the blockchain mainly refers to the definition and attribution of data sovereignty caused by different subjects involved in data generation, transmission, storage, analysis, processing, and application in the era of big data. For example, data sovereignty on social networking sites involves multiple profit holders, including user groups, service providers of Internet communication,

Internet access, data centers, and social networking sites. Pursuant to the laws and regulations of China, China may have the power to restrict the exit of data. So, to whom the right to data use and ownership of data belong, individuals, enterprises, or nations? How should the boundaries of all parties be determined?

Clear data ownership is the basis of dealing with the above questions. Through data ownership confirmation, the relationship between rights and obligations and responsibility boundaries among the subjects of the whole life cycle of data can be clarified, so that behaviors such as data collection, storage, transmission, use, disclosure, and transaction can be expected, and the sustainable development of the data industry is effectively guaranteed. "Generally speaking, data ownership confirmation determines the rights holder of the data, that is, who has the ownership, possession, use and earning rights of the data, as well as the responsibility to protect personal privacy rights, etc." (Hailong et al. 2018). It aims at balancing the interests of all data parties, so the interests of all parties are protected, and the data can be widely used as well as explored in the greatest value. Starting from the goal of data ownership confirmation, current data ownership confirmation is mainly to address data rights subjects, data rights attributes, and data rights contents. To be specific, it includes who should enjoy the benefits of the data, what kind of rights protection should be given to the data, and what specific power the data subjects enjoy.

Data ownership confirmation has become the key to protecting personal privacy, promoting industry, and ensuring national security. However, data ownership confirmation seems simple at present but is particularly complicated. On the one hand, data sources are extensive, and there are significant differences in the perception and focus of data ownership among individuals, companies, and nations. On the other hand, different factors, e.g., data analysis ability, data technology level, and data control ability impact data ownership confirmation (Jiang 2018). From this perspective, it is necessary to clarify the boundaries of three levels, namely, "the open boundary of national data, the commercial application boundary of enterprise data and the privacy protection boundary of personal data" (Wei and Changyue 2017) in order to have a clear definition of data sovereignty.

2.8.3 Data Sovereignty Under Blockchain

In the era of big data, the Internet information network fails to confirm data ownership although it facilitates the sharing of data. Meanwhile, the traditional method of data ownership confirmation adopts the mode of submitting ownership certificates and expert reviews, which lack technical credibility and maybe in the danger of uncontrollable factors, e.g., potential tampering. For tackling these problems, there is an urgent need for a feasible data ownership confirmation plan. In response, people propose a new data ownership confirmation plan based on blockchain technology. As an emerging information technology which is progressing rapidly in recent years, blockchain realizes distributed ledger without a need for trust in a suspicious environment. The natural characteristics of the distributed ledger, such as traceability, high reliability, immutability, transparency, trustworthiness, and decentralization, give the blockchain unique advantages in data ownership confirmation, thus effectively tackling "data ownership confirmation". First, "miners" mark the data with "time stamp" to distinguish the data transmitted before and after. Second, with "Smart Contract", the flow of property rights is to be realized while data is transmitted among different subjects. Third, with the "distributed ledger", the mechanism of mutual supervision and restriction by multiple subjects, the whole process can be realized.

Born as the underlying technology of Bitcoin, the blockchain technology is actually a set of data systems with very high security. Data is recorded through encryption algorithm and chain data structure. If the data of a certain block is attacked and tampered with, it is impossible to obtain the same Hash Function, and it can be easily identified by other network nodes, so the integrity, tamper-resistance, and uniqueness of the data can be ensured. Compared with traditional centralized data storage, blockchain technology adopts a new data storage mode to maintain the data sovereignty of data subjects. First of all, the blockchain technology has changed "the naming, indexing and routing modes of data objects in the current Internet. Decentralized digital object management and routing decouple data and applications thus supporting efficient and credible data sharing and releasing development space for highly integrated applications" (Xiaolan 2019). Second, the blockchain technology changes the communication and computing mode of the Internet which is "supplemented by computing and storage, and centered by network communication" into a mode supplemented by communication based on private data centers, and centered by computing and storage, so data subjects can have total control over data and realize the aim "I take charge of my own data".[23]

Blockchain is a new type of production relations in the information society. It was born to address cooperative bookkeeping among multiple data subjects, which is also the core value of blockchain. The blockchain realizes not only "I take charge of my own data", but "We witness our history together". On the one hand, the tamper-resistance of blockchain makes data difficult to be tampered with without permission, thus achieving mutual trust among participants. On the other hand, blockchain, together with cryptology technology, protects users' data privacy more, achieving win–win cooperation among multiple parties in the game. On the blockchain, one data subject must be authorized by another subject to see the latter's data, so the blockchain helps protect data sovereignty. No matter how small the data subject is, it can make decisions for its own data and safeguard data dignity. Meanwhile, as an accounting technology based on distributed ledger technology, blockchain facilitates many to keep accounts together, thus achieving "we witness our history together".

---

[23]Yin Hao, Li Yan. Promoting Blockchain and Safeguarding Internet Data Sovereignty. Chinese Institute of Electronics. 2019. https://www.btb8.com/blockchain/1906/55993.html.

2.9 Data Sovereignty Game

Data sovereignty has become the focus of the game between all parties, and there has been intensifying competition among nations for data dominance. In practice, the absolute independence of data sovereignty leads to multiple jurisdiction conflicts and national security dilemmas. The confrontation and game of data sovereignty result in a disorderly international community in data space. Under this background, for tackling this kind of dilemma, all nations shall resume the cooperative participation of sovereignty, transfer and share some data sovereignty, and safeguard data sovereignty by ensuring data security and strengthening data governance, for its sound development, lasting peace, universal security, and common development of humanity.

2.10 Confrontation of Data Sovereignty

The cross-border flow and storage of data transcend traditional absolute independent sovereignty theory. An independent sovereignty nation can neither exercise complete and autonomous possession and jurisdiction over its own data nor exclude external interference completely. If a nation exercises absolute unilateral control over data and related technologies on the grounds of data sovereignty independence, it will lead to spontaneous confrontation and game of data sovereignty and eventually to a disorderly international community in data space. Thus, the over-emphasis on the independence of data sovereignty is the main factor that leads to the confrontation of data sovereignty among nations. Based on data sovereignty independence at present, the spontaneous game of data sovereignty mainly originates from multiple jurisdictional conflicts over data and the dilemma of national data security.

Conflicts of Multiple Jurisdictions over Data. The big data era poses multiple challenges to international law, among which the governance of data in different nations based on the different places of possession, storage, or transmission exerts the most lasting and profound impact. Meanwhile, to reduce costs and cater to customers, data service providers often outsource some of the services. Therefore, the same data is likely to be subject to multiple jurisdictions of different nations, especially at present, nations have not yet defined the jurisdiction scope of data sovereignty, and nations exercise their rights in the international community in a completely rational way. Before the formation of an international unified system or coordination mechanism, for ensuring the absolute security of the nation and for monitoring, all nations propose data sovereignty over all the data that can be monitored, which will inevitably lead to the monitoring of some extraterritorial data, thus to multiple jurisdictions. In addition, given the free flow of data among nations, it is inevitable that there will be overlapping jurisdictions between data and related subjects. If all nations advocate data sovereignty that is beneficial to themselves without any sacrifice or concession, there will inevitably be conflicts and confrontations of data sovereignty among nations.

The Dilemma of National Data Security. First of all, compared with developed countries, both developing countries and under-developed countries have insufficient control over data. Although they have independent data sovereignty, many developing countries and under-developed countries fail to effectively safeguard their own data security and national interests due to the limited level of data technology. Second, the data technology revolution not only makes some developed countries abuse data sovereignty by taking advantage of their technological advantages but also threatens the data security of other nations. The United States, for example, not only exercises control rights over extraterritorial data related to America through *USA PATRIOT Act* but also collects and analyzes data under the jurisdiction of other nations through special projects of the National Security Department. The "Prism" incident evidenced that the US security department stole data and information from other nations. Snowden disclosed to the media in 2013 that through the "Prism" project, the US Government collected information directly from the servers of nine companies, including Microsoft, Google, and Yahoo. It stole the user data of all mainstream smartphones, including Apple's mobile phones, covering e-mail, communication information, and Internet searches. Meanwhile, the use of spyware and encryption technology for monitoring by the United States has also been much reported. The frequent exposure of similar incidents reflects the serious threat posed by America, a network power, to the data security of other nations. In addition, the spontaneous game of data sovereignty also makes it difficult to effectively guarantee national data security. On the one hand, the cross-border flow and storage of data greatly weakened a nation's effective jurisdiction over data and related equipment, forming serious security bugs. On the other hand, relying on their technological advantages, developed countries can collect and monitor the data of other nations in some hidden way, thus infringing on the data sovereignty of other nations. Therefore, the emphasis on the independence of data sovereignty leads to confrontation between nations, so some developed countries will arbitrarily implement unilateralism in data space (Nanxiang and Xiaojun 2015).

Spontaneous Game of Data Sovereignty. The independence of data sovereignty is closely related to the conflicts of multiple jurisdictions over data and national data security dilemma, forming the confrontation of spontaneous game of data sovereignty. First of all, the emphasis on the absolute independence of data sovereignty will lead to conflicts of multiple jurisdictions over data. Data flow involves at least data producers, recipients and users, the place of data transmission, transportation and destination, the place of data infrastructure, nationality, and business location of data service providers, etc. Due to the indivisibility and integrity of data, cross-border data behaviors in any aspect will lead to the overlapping of national jurisdiction and conflicts of data sovereignty. Meanwhile, in the case of multiple jurisdictions over data, service providers are likely to select particular laws to their advantage, to help them evade domestic regulations on data protection through data transfer, thus affecting the data security of the nation. Second, in terms of data security, a nation will take absolute unilateral control over data and related technologies, taking data sovereignty independence as an excuse, especially impose legal restrictions on the location of data centers, to be established within the security control scope designated by the nation. This actually prohibits potential foreign data service providers from providing services to customers, giving big data a boundary and undermining cost advantage, the basis on which data technology develops. Finally, the network powers headed by the United States will obtain sensitive data by infringing on the data sovereignty of other nations on the grounds of exercising data sovereignty. For example, the United States monopolizes the world's Internet root server resources and has a large number of most influential communication service providers and network operators in the world. Therefore, it is apt to steal the secret data of other nations, which threatens global data security.

With the rapid development and popularization of data technology, different nations define data jurisdiction from various angles. In particular, based on the independence of data sovereignty, all nations have claimed jurisdiction over extraterritorial data. Meanwhile, the big data era has weakened a nation's control over its own relevant data. Under data sovereignty confrontation, many developing countries and under-developed countries are not able to ensure their data security or effectively exercise data sovereignty without the aid of international coordination mechanisms. However, developed countries can effectively exercise their data sovereignty with advanced technologies, and even endanger the security of other nations' data sovereignty. From this point of view, the absolute independence of data sovereignty leads to the conflict of multiple jurisdictions over data and national data security dilemma. The spontaneous confrontation and game eventually lead to the disorder of the international community in the data space. Therefore, for rectifying the current disorder of data space, we should explore the transfer cooperation based on data sovereignty and establish corresponding international coordination organizations or mechanisms.

2.11 Transfer and Sharing of Data Sovereignty

In the early days of sovereignty theory, most believed that sovereignty was absolute, permanent, indivisible, and inalienable. This view existed before eighteenth century, represented by the following figures. First, Bodin and his modern state sovereignty theory. Bodin believed sovereignty, an indivisible and transferable permanent power, appeared as a ruling force divorced from and above the society. The government may change but sovereignty exists forever. Second, Hobbes and the theory of monarchical sovereignty. Hobbes maintained that sovereignty should be absolute, infinite, as well as indivisible and inalienable. He pointed out "the so-called statement which claims sovereignty can be divided directly violates the essence of the nation. To divide the rights of the nation is to disintegrate the nation, because the divided sovereignty will destroy each other" (Hobbes 1986). Third, Locke and his theory of parliamentary sovereignty. Locke held that the legislative power was the supreme power in the society, and the legislature could not transfer legislative power. When the community transfers legislative power to the legislature, this supreme power is sacred and unchangeable. Fourth, Rousseau and his theory of people's sovereignty. Rousseau believed that the essence of sovereignty was a common will (public will) of all the people. Sovereignty is indivisible and inalienable, so it is a whole. Any division will turn the public will into the will of individuals, thus putting sovereignty out of existence (Rousseau 2003). From these theories, it can be seen that in the traditional view of sovereignty, sovereignty is sacred, i.e., it must be indivisible and inalienable.

The sovereignty theory is facing new challenges in the era of data globalization. The traditional sovereignty theory which emphasizes absoluteness, indivisibility, and inalienability no longer meets the development of the times. As Bertrand Bhatti, an American scholar pointed out in *Globalization and Open Society*, "Globalization destroys sovereignty states, connects the world map, abuses the political community it has established, challenges social contracts, and prematurely puts forward useless national guarantees… Since then, sovereignty is no longer an indisputable basic value as it used to be". The adherence to the absoluteness, indivisibility, and inalienability of national sovereignty is out of steps with the times. In the new era, national behaviors are more and more restricted by other actors. Facing the restricted reality of sovereignty, a new trend of thought that sovereignty can be transferred begins to emerge. It is generally believed sovereignty transfer meant sovereignty states voluntarily transferred part of their sovereignty powers to other nations or international organizations on the basis of the principle of sovereignty and reserved a way of exercising sovereignty by withdrawing part of the transferred sovereignty powers at any time, in the background of globalization, based on the division of identity sovereignty and power sovereignty, in order to maximize national interests and promote benign interaction and international cooperation among nations.

"Sovereignty transfer comes out of the collision between globalization and national sovereignty, which has its rationality and inevitability. Therefore, most people hold a positive attitude towards sovereignty transfer" (Fei 2009). However, the concept of sovereignty transfer has been in dispute since it came into being. The direct cause is that scholars use different elements of national sovereignty to refer to national sovereignty, and the dispute is rooted in the multi-element connotation of national sovereignty. State sovereignty contains different elements such as sovereignty identity, sovereignty authority, sovereignty power, sovereignty will, and sovereignty interests. There are different answers to the question of whether these elements can be transferred. Elements such as sovereignty power and sovereignty interests can be transferred, while elements such as sovereignty identity, sovereignty authority, and sovereignty are non-transferable. There is no consensus currently on the word "sovereignty transfer", and scholars' application of the concept "sovereignty transfer" is rather chaotic, such as "sovereignty shift" and "sovereignty transfer". From the standpoint of developing countries, most Chinese scholars, when demonstrating the relation between sovereignty nations in globalization, believe that alienation is an active, positive, autonomous, and voluntary act, a legal act, and a realistic state, indicating the alienation and transfer of ownership. It does not mean the conceded third party will completely lose the transferred power after transferring it, but it is a positive act (Shanwu 2006).

The theory of sovereignty transfer in the era of big data has been accepted by more and more people and has been endowed with new connotations. It is not only the transfer of national sovereignty but also a deep-level transfer including individual sovereignty and enterprise sovereignty. As an extension and expansion of sovereignty in the era of big data, like national sovereignty, data sovereignty can be transferred from one data subject to another without endangering national security or damaging the legitimate interests of enterprises and individuals. In other words, data sovereignty is divisible and transferable, but this transferability is only partial and temporary, and can be fully recovered by the data subject after an independent decision. The transfer of data sovereignty is the result of the data subjects exercising data sovereignty, which embodies the necessary response of data sovereignty in facing challenges. Data sovereignty transfer is sharing, autonomous, and free. Sharing means achievement sharing, and no data subject is allowed to have privileges within a specific range. Autonomy and freedom refer to the fact that each data subject is voluntary in joining or quitting, and is not subject to any coercion or restriction. The transfer of data sovereignty is not to give up data sovereignty but to share data sovereignty to maximize the interests of communities and individuals (Yikang and Haibing 2003). In data sovereignty transfer, "the data sovereignty of enterprises and individuals must unconditionally obey the needs of the national data sovereignty, as national data sovereignty is the most important" (Lin and Kexi 2016).

The conjunction of personal, enterprise, and national interests provides space for the transfer of data sovereignty, which is also a necessary condition. However, in a theoretical and practical level, there must be contradictions among personal, enterprise, and national interests. Under such circumstances, the transfer of data sovereignty will inevitably be affected or even hindered by many factors. Of course, the contradiction between personal, enterprise, and national interests has relative influence and obstacles on the transfer of data sovereignty; otherwise, data sovereignty transfer in the era of big data would not be common. The biggest reason lies in that data sovereignty transfer provides a better choice and path for the harmonious development of individuals, enterprises, and nations. It is not only a process of coordinating all subjects for fair distribution of benefits and multi-win games but also a result of data sovereignty cooperation among all parties involved. "Repeated games on the expected benefits of cooperation, collision and shock of interests of all parties, finally it comes to the balance of interests" (Kai 2007).

In the international community, the transfer of data sovereignty is mainly reflected at the national level, as it must adhere to the principle of national interests. Hans Morgenthau, a famous American scholar, pointed out in *The Dilemma of Political Science*, "As long as the world is still made up of nations politically, the final language in international politics can only be the national interest". This shows that the national interest sets the basic objectives of a nation's foreign policy and determines the law of a nation's international behavior. In the era of big data, whether and how the nation transfers data sovereignty in the data space depends on national interest. Only by adhering to national interests can the state partially transfer its data sovereignty. The transfer of data sovereignty "sacrifices temporary and local interests to some extent in exchange for long-term interests and overall interests". In addition, a nation's transfer of data sovereignty is not unlimited, unprincipled, or is carried out passively under the threat by power. There is also an issue called "degree" in the transfer and autonomy restriction of data sovereignty, that is, in national data sovereignty transfer, it is necessary to ensure the independence and autonomy of the nation, as well as the equality of rights and status among nations.

2.12 Data Sovereignty, Data Security and Data Governance

Keeping data sovereignty is of great significance to national security and development, but data sovereignty faces new threats and challenges in practice, such as data security, data hegemonism, data protectionism, data capitalism, and data terrorism. Therefore, on the principle of data sovereignty, it is urgent to construct a data sovereignty legal system that adapts to the current development situation, to speed up data security legislation and improve the data governance system, so as to reduce the abuse of data sovereignty and ensure its sound development.

2.12.1 Challenges and Countermeasures of Data Sovereignty

With the progress of data globalization, data sovereignty is facing severe challenges. On the one hand, different nations are unable to effectively exercise data sovereignty, and their capacity to store and control data is weakened correspondingly due to the different legislative modes and strategies adopted by different nations for data management and protection, plus the cross-border flow of data, the characteristics of data processing and the data sovereignty game between nations. On the other hand, as the international community has not clearly defined data sovereignty, data sovereignty has yet to be established in international law, and there are a lot of problems in claiming data sovereignty among nations. Meanwhile, as a new national right, data sovereignty is facing new challenges and threats including data security, data hegemonism, data protectionism, data capitalism, and data terrorism now. Therefore, establishing the basic principle of data sovereignty as soon as possible, keeping up with international legislation, and constructing a data sovereignty system that caters to the current development situation will be more conducive to safeguarding the sovereignty and stability of all nations.

The System Construction of Data Sovereignty. The relevant laws and policies of data sovereignty at present are mainly carried out on the management and control of data, while the claims and practices of various nations on data sovereignty are mainly manifested in the management demands of cross-border data flow. Globally, more and more nations start to construct their data sovereignty-related systems around data management from the legal perspective and show three development trends.

First, in order to maintain their national data security, restrictions are imposed on the cross-border export of important data; second, to strengthen data control, they make legislative adjustments to the localized storage of personal data; third, they extend extraterritorial jurisdiction over data (Bo 2017). The legal system of data sovereignty should not only highlight the security and protection of core data; moreover, the mining and utilization of big data resources, as well as the economic losses and technical risks caused by data hegemony. Therefore, relevant regulations and systems should be constructed from three levels; namely, localized storage of data resources, hegemony resolution in the dimension of community of shared future, and cross-border flow based on data classification (Jianwen and Zhangfan 2018).

The Legislative Direction of Data Sovereignty. "When legislating data sovereignty, we should analyze it from various aspects in terms of a multi-dimensional perspective. National data sovereignty security is a diversified, multilateral and democratic comprehensive system" (Aimin and Gaofeng 2016). From the perspective of domestic laws, we should improve the relevant laws and regulations on data sovereignty, speed up the legal stipulation of data sovereignty status, strive to outline the specific framework of data sovereignty regulation system, exercise data sovereignty within legal framework, and protect data security and national interests. Meanwhile, we should make full use of the cooperation with other nations in various fields, draw lessons from the advanced data protection experience of Europe and America. In light of the fact that data sovereignty legislation is not yet perfect, we should establish a systematic legal framework of data rights in combination with actual national conditions, improve data review mechanism, and legislation technologies in the data field. From the perspective of international laws, all nations should be participants in the formulation of international rules on data security and relevant data security treaties based on the principle of "seeking common ground while reserving differences". The treaty is intended primarily to guide the international community in achieving data security, to guide nations for more international cooperation, to crack down on data abuse, infringement, and damage to other nations' data sovereignty, eliminate non-security risks that threaten other nations' political, economic and social data security, to ensure fair and just distribution of data resources, and maintain data security, stability, and free operation (Aimin and Gaofeng 2016).

2.12.2 The Essence of Data Security: National Data Sovereignty

When data has become a national basic strategic resource and a social basic factor of production today, the biggest problem is data security, and military, political, economic, cultural, sci-tech, and social security based on data security. Whoever masters data security will occupy the commanding heights of the big data era and have "data dominance". Data security has become the top priority of national security in the era of big data, and there is a closer relationship with other security factors.

It has risen to an overall strategic position that affects the nation's political stability, social stability, and orderly economic development, so it is the cornerstone of national security.

In essence, data security is a nation's data sovereignty. In a sense, data sovereignty will be lost without data security. Data security involves military, political, economic, and cultural fields. As big data is unbalanced and insufficient in geographical distribution, there has been a "data potential difference" between strong and weak data powers at the strategic level. Nations with low data potential will face unprecedented serious threats and challenges to their security in political, economic, military, and cultural fields, while big data has become a new tool for strong data powers to seek future strategic advantages. The "data territory" is divided based on data radiation space with some political impact, but not on territorial waters, land, and sky of sovereignty states, or on geographical characteristics. The security of data boundaries, the mastery of data sovereignty, and the size of "data territory" are the keys to the rise and fall of nations in the era of big data (Jianmin 2020). Safeguarding data sovereignty and ensuring data security have become major challenges to governments in the big data era. Therefore, data sovereignty and data security are a pair of inseparable concepts, and there are countless interactions and dialectical relationships between them.

Legislation on data security is the shield to data sovereignty. Legislation should come first in ensuring data security. Only in the rule of law can a sound balance between free data circulation and the supervision of cross-border data transfer be realized, and the safety of the nation, public interests, and individuals can be ensured in the flow and application of data. However, there lack unified national legislations on data security in China so far. Relevant regulations, only read in different laws and regulations, fail to provide effective legal support for data sharing and for preventing data abuse and infringement. Therefore, it is urgent to take targeted legal measures to build a legal and regulatory system of data security. *Regulations of Guiyang City on Big Data Security Management*, the first local legislation in the field of data security in China, drafted by Professor Lian Yuming, director of the Key Laboratory of Big Data Strategy, provides valuable and replicable experience for data security legislation at the national level. Nowadays, with mature conditions for national protection of data security by laws, there has been a stronger voice for the legislation on data security. In the NPC and CPPCC sessions, many NPC deputies and CPPCC members strongly called for data security legislation at the national level. Among them, Lian Yuming submitted *Proposal on Accelerating Data Security Legislation* in March 2018 and *Proposal on Accelerating the Legislative Process of "Data Security Law"* in March 2019. In September 2018, the *Data Security Law* was officially listed into the legislative plan of the 13th National People's Congress Standing Committee, and data security legislation elevated to national legislation from a local one.

2.12.3 Data Governance Rises to National Strategy

Data governance came to be known first in enterprise-wide procedures (Begg and Caira 2012). Cohen defines data governance as "the amount, consistency, availability, security and controllability of data managed by companies", or the collection of policies, processes, standards, decisions, and decision-making rights. Having no will or self-intention, data is shaped as a means of the instrument and is told where to be oriented, so data needs to be controlled. "Meanwhile, governance itself is a technical project" (Yaqian and Gang 2018). It can be said that in early research, data governance is the governance between humans and technology (Coleman 2008). With the advancement of research, the connotation of data governance has been enriched, and emphasis has been placed on identifying roles or organizations with data authority. Data governance comes to be applied to nations, and more nations have begun to elevate data governance to the strategic level. Data governance means a nation applies relevant laws, regulations, and principles to individuals and enterprises that violate regulations, so as to achieve harm-free, standardized, and legalized data production and application, to realize controllable and available data from the national perspective as a whole, and control data generation, circulation, and use, to prevent the damage caused by data development to ruling security, ideological security, and data sovereignty security.

From the Internet which connects people, to the Internet of Things and Industrial Internet which connects objects and objects, to the new information technology represented by 5G technology which promotes the Internet of Everything, explosive data volume provides a major prop for sci-tech innovation. However, data governance still faces problems and challenges in terms of data security needs and freedom anxiety, expansion and lack of data supervision, the excessive and insufficient protection of personal data, and competition among sovereignty nations. For example, different nations have different understandings in protecting citizens' privacy rights and national data security, which makes it difficult for nations to unify the two in the global scope. For another example, there are more and more international disputes on data governance in the context of rising unilateralism. Data localization and data review aggravate the fragmentation of data governance. In addition, the maintenance of data sovereignty is another challenge facing data governance. Data sovereignty is subject to the attributes and characteristics of data technology itself, which is different from judicial, diplomatic, and territorial sovereignty. Data governance meets challenges in characteristics, boundaries, connotation, and countermeasures (Rui 2019).

In the context of data sovereignty game, it is unrealistic to engage in data governance behind closed doors. The spillover effect and externality of the rules of the game must be taken into account. Therefore, in order to legalize data governance, we need to upgrade the interaction between the domestic rule of law and the international rule of law and establish a global data governance system that not only protects national interests but makes dialogue, competition, and cooperation possible, so as to raise China's voice and governance capability in the global data governance system

(Xixin 2019). At present, America and Europe have established a relatively complete data governance system with the aid of the CLOUD Act,[24] GDPR and other legal rules, and the system includes personal privacy protection, data sovereignty, and cross-border data flow. "In contrast, personal information protection in our nation is still relatively weak; there is a risk of endangering national security due to personal information disclosure, and the data governance system is not perfect enough" (Xiao and Xiaoyu 2019). On December 8, 2017, when presiding over the second collective study of the Political Bureau of the CPC Central Committee on the implementation of national big data strategy, General Secretary Xi Jinping stressed "we should build up our reserve of international data governance policies and conduct more research on governance rules, and propose China's plans". As the world's largest data nation, we should make full use of our unique advantages, e.g., data size and scenario application, take countermeasures from the perspective of system, law, and rules, step up a national data governance rule system, and "combat the wanton retrieval of China's data by foreign governments", (Shuyin 2018) so as to ensure the sound development of network power, digital China and a smart society.

References

I. Chinese Monographs

Key Laboratory of Big Data Strategy. *Data Rights Law 1.0: Theoretical Basis of Digital Power*. Social Sciences Academic Press, 2018, editor's note.

Duan Fan. *Power and Rights: Co-ownership and Construction*. People's Publishing House. 2016.

Li Buyun. *Jurisprudential Analysis*. Hunan People's Publishing House, 2003.

Qi Yanping. *Evolution of the Ideology of Human Rights*. Shandong University Press, 2015.

Wang Guanghui. *Human Rights Law*. Tsinghua University Press, 2015.

Xu Chongde. *Constitutional Law*. China Renmin University Press, 2009, p.196.

Zhang Wenxian. *Research on Western Philosophy of Law in the 20th Century*. Law Press, China, 2006.

Zhang Wenxian. *Jurisprudence (4th Edition)*. Higher Education Press, Peking University Press, 2011.

Zhao Zhou. *On Sovereign Responsibility*. Law Press, China, 2010.

Zhou Gengsheng. *International Law (Volume I)*. Wuhan University Press, 2009.

Zhuo Zeyuan. *On the Country Ruled by Law*. China Fangzheng Press, 2001.

[Germany] Jürgen Habermas. *Faktizität und Geltung*, translated by Tong Shijun. SDX Joint Publishing Company, 2003.

[Germany] Kant. *Metaphysical Principles of Law--Science of Rights*, translated by Shen Shuping. The Commercial Press, 1991.

---

[24]CLOUD Act, Clarify Lawful Overseas Use of Data Act passed by the US Congress in March 2018, adopts the so-called "data controller standard" to make it clear that the power of US Law enforcement agencies to retrieve data from network operators have extraterritorial effects and are attached to the corresponding principles of international comity. Meanwhile, it sets up a mechanism for foreign governments to retrieve data from the United States.

[France] Jean-Jacques Rousseau. *Du Contrat Social*, translated by He Zhaowu. The Commercial Press, 2003.

[Ancient Rome] Cicero. *De Re Publica; De Legibus*, translated by Shen Shuping, Su Li. The Commercial Press, 1999.

[Ancient Greece] Plato. *The Republic*, translated by Guo Bin, Zhang Zhuming. The Commercial Press, 1986.

[Ancient Greece] Aristotle. *The Politics*, translated by Wu Shoupeng. The Commercial Press, 1965.

[US] J. Feinberg. *Freedom, Rights and Social Justice*, translated by Wang Shouchang et al. Guizhou People's Publishing House, 1998.

[US] Peter Demandis, [US] Stephen Kotler. *Abundance: The Future Is Better Than You Think*, translated by Jia Yongmin. Zhejiang People's Publishing House, 2014.

[US] Dennis H. Wrong. *Power: Its Forms, Bases, and Uses*, translated by Lu Zhenlun, Zheng Mingzhe. China Social Sciences Press, 2001.

[US] Barker, Kurt W. *Social Psychology*, translated by Department of Sociology. Nankai University Press, 1984.

[Japan] Osuga Akira. *The Right of Existence*, translated by Lin Hao. Law Press, 2001.

[UK] A.J.M. Milne. *Human Rights and Human Diversity--An Essay in the Philosophy of Human Rights*, translated by Xia Yong, Zhang Zhiming. Encyclopedia of China Publishing House, 1995

[UK] Thomas Hobbes. *Leviathan*, translated by Li Sifu, Li Yanbi. The Commercial Press, 1986.

[UK] Jennings, [UK] Watts revised. *Oppenheim's International Law (Volume 1, Book1)*, translated by Wang Tieya et al. Encyclopedia of China Publishing House, 1995.

II. Chinese Journals

Zeng Huan. On the Dialectical and Unified Relation between Human Rights and National Sovereignty. *Legal System and Society*, 2015, No. 5.

Chen Xiuping, Chen Jixiong. Balance of Public Power and Private Rights from the Perspective of Rule of Law. *Seeker*, 2013, No. 10.

Chen Zhiying. Reconsideration of Modernity of Sovereignty and Its Return to Commonality. *Modern Law Science*, 2007, No. 5.

Dou Yanguo. On Public Power and Citizen Right. *Research on Mao Zedong and Deng Xiaoping Theories*, 2006, No. 5.

Feng Wei, Mei Yue. In the Era of Big Data, Data Sovereignty Dominates Ups and Downs. *Information Security and Communications Privacy*, 2015, No. 6.

Fu Wei, Yu Changyue. A Review of Research and Dynamic Analysis of Development on Data Rights at Home and Abroad. *Journal of Modern Information*, 2017, No. 7.

Guo Daohui. The Particularity and Gist of Power. *Journal of Shandong University of Science & Technology (Social Sciences)*, 2006, No. 2.

He Bo. Research on Legal Practice and Countermeasures of Data Sovereignty. *Information Security and Communications Privacy*, 2017, No. 5.

He Zhe. Forms and Orders of Human Society in Network Civilization. *Nanjing Journal of Social Sciences*, 2017, No. 4.

He Tianping, Song Wenting. Historical Evolution of "Number-Data-Big Data". *Studies in Dialectics of Nature*, 2016, No. 6.

Jiang Jiang. The Ownership Structure and Confirmation of Data. *New Economy Weekly*, 2018, No. 7.

Jiang Guangning. Public Power and Private Rights in a Law-based Country. *Knowledge Economy*, 2010, No. 24.

Li Aijun. Attributes and Legal Characteristics of Data Rights. *Oriental Law*, 2018, No. 3.

Li Xiao, Gao Xiaoyu. Heeding the Trend of International Data Governance Game and Safeguarding China's Data Sovereignty. *Secrecy Science And Technology*, 2019, No. 3.

Liu Hong, Hu Xinhe. Data Revolution: Historical Exploration from Number to Big Data. *Journal of Dialectics of Nature*, 2013, No. 12.

Liu Kai. An Analysis of the Difficulties and Problems Restricting National Sovereignty Transfer in the Era of Globalization. *Theory and Modernization*, 2007, No. 3.

Liu Xiaochun, Wu Qiong. Alienation and Control of Public Power. *Reform & Opening*, 2012, No. 10.

Long Rongyuan, Yang Guanhua. The Research on Data Right, Data Right System and Data Right Law. *Science Technology and Law*, 2018, No. 5.

Ma Changshan. The Fourth Generation of Human Rights and their Protection in the Context of a Smart Society. *China Legal Science*, 2019, No. 5.

Ni Jianmin. The Development of Informatization and the Information Safety of Our Country. *Journal of Tsinghua University (Philosophy and Social Sciences)*, 2020, No. 4.

Pan Aiguo. On the Boundary of Public Power. *Jin Ling Law Review*, 2011, No. 1.

Qi Aimin, Zhu Gaofeng. On the Establishment and Improvement of National Data Sovereignty System. *Journal of Soochow University (Philosophy & Social Science Edition)*, 2016, No. 1.

Shi Dan. The Legal Protection and Construction of Property Rights in Enterprise Data. *Electronics Intellectual Property*, 2019, No. 6.

Sun Nanxiang, Zhang Xiaojun. On Data Sovereignty—An Examination of Game and Cooperation in Cyberspace. *Pacific Journal*, 2015, No. 2.

Tao Lin. The Tension and Agreement between Human Rights and Sovereignty. *Philosophical Researches*, 2013, No. 5.

Wang Hailong, Tian Youliang, Yin Xin. Blockchain-based Big Data Right Confirmation Scheme. *Computer Science*, 2018, No. 2.

Wang Lin, Zhu Kexi. The New Perspective on Data Sovereignty. *Journal of Yunnan Agricultural University (Social Science)*, 2016, No. 6.

Wei Shuyin. CLOUD Act Implies US Data Hegemony Attempt. *China Information Security*, 2018, No. 4.

Wu Yikang, Zhang Haibing. On the Transfer of Sovereignty--A Debate on the Article "On the 'Indivisibility' of Sovereignty". *Chinese Journal of European Studies*, 2003, No. 6.

Xie Tao. The Game between Public Power and Private Rights. *Knowledge Economy*, 2011, No. 21.

Xu Wei. Reflection on the "Triple Authorization Principle" of Enterprise Data Acquisition and Typological Construction. *SJTU Law Review*, 2019, No. 4.

Xu Xiaolan. Research on Blockchain Technology and Development. *Electronic Technology & Software Engineering*, 2019, No. 16.

Xu Yaqian, Wang Gang. Data Governance Research: Process and Contention. *E-Government*, 2018, No.8.

Yang Fei. An Analysis of the Definition of the Concept of State Sovereignty Transfer. *Journal of University of International Relations*, 2009, No. 2.

Yi Shanwu. A New View on Sovereign Transfer. *Journal of Chongqing Jiaotong University (Social Sciences Edition)*, 2006, No. 3.

Yu Zhigang. The Right Attribute of "Citizen's Personal Information" and the Thought of the Criminal Law Protection. *Zhejiang Social Sciences*, 2017, No.10.

Zhang Jianwen, Jia Zhangfan. Interpretation Logic and System Construction of Data Sovereignty from the Perspective of Law and Economics. *Journal of Chongqing University of Posts and Telecommunications (Social Science Edition)*, 2018, No. 6.

Zhang Wenxian. Jurisprudence of Human Rights in a New Era. *Human Rights*, 2019, No. 3.

Zhao Gang, Wang Shuai, Wang Peng. Study on Data Sovereignty Oriented Technical Schema for Big Data Governance. *Cyberspace Security*, 2017, No. 2.

III. Chinese Newspapers

Lin Zhe. What are Human Rights? *Study Times*, March 1, 2004, p.T00.

Qiu Rui. "The Governance of Data" Promotes "The Governance of China". *Study Times*, December 27, 2019, p.7.

Wang Xixin. Data Governance Legislation Can Not Neglect the Principle of Rule of Law. *Economic Information Daily*, July 24, 2019, p. 8.

V. Monographs in Foreign Languages and their Excerpts

Coleman S. "Foundations of digital government"//Chen H. *Digital Government*. Boston, MA: Springer, 2008, pp. 3-19.

VI. Foreign Journals

Begg C, Caira T. Exploring the SME quandary: Data governance in practice in the small to medium-sized enterprise sector. *The Electronic Journal Information Systems Evaluation*, 2012.

Chapter 3
Social Trust Theory

Trade cannot exist without trust, and it is very difficult to trust strangers.

—Yuval Noah Harari, an Israeli historian.

Blockchain is a machine of trust in a world "out of control".

—Kevin Kelly, Founding Editor-in-Chief of Wired.

The history of mankind follows the logic of division and unity in turn, and blockchain technology is taking the Internet era into such a logic. We are now witnessing a new revolution in this era brought about by blockchain and decentralized technology.

—Shou-Cheng Zhang, a Renowned Chinese American physicist.

3.1 Trust and Consensus

Trust and consensus are the crucial concepts of blockchain. Trust, one of the main elements of social capital, is a conscious act of an individual, an essential form of social consensus relationship, and also the prerequisite and foundation for economic operation and social stability. Consensus is a set of routines and customs shared by general public, which is both a goal and an institution and has a bearing on the achievement of specific policies. Initially, consensus was only a concept in society operation, but now it has been an important part of computer science. Whereas the biggest problem of the Internet is its incapacity to address the problem of trust, blockchain provides a solution that goes completely beyond conventional thinking. Blockchain creates a consensus mechanism based on consensus and mutual trust without a trusted central node and thus sets a decentralized trustworthy system without having to trust a single node, which marks a fundamental move from decentralized state credit toward centralized algorithmic credit.

© Zhejiang University Press 2021
L. Yuming, *Sovereignty Blockchain 1.0*,
https://doi.org/10.1007/978-981-16-0757-8_3

3.2 Trust and Social Order

Trust is the lubricant of social systems. As a phenomenon of social structure and cultural norm, trust is a behavioral expectation formed beyond the known information and a "simplification mechanism" to lessen the complexity of social interaction. The human society is unique, in that trust permeates all human interactions and contains both lofty aspirations and deep-seated fears. Mutual trust between people based on rational cognition and certain principles is a safeguard for social order.

3.2.1 Trust and Reliance in an Acquaintance Society

Human society is constituted by kinship for most of the time in history. Different social organizations have different forms of trust. Personality trust in an acquaintance society is the most basic type of trust to maintain social order and regulate social behavior through information sharing in the ideology of collective maintenance. Such personality trust featuring intimate relationship and full trust (Guoqiang 2020) satisfies the communication in an acquaintance society with clear interpersonal boundaries (Hong 2011). The unique traditional inheriting system, kinship system, Confucian ethics, and morals of superiority and inferiority, as well as the characteristics of farming culture of living in a compact community, have formed the ideological framework for a Chinese acquaintance society, where trust is closely linked to the Differential Mode of Association.[1] In such a society where people are born and bred in the same place, the culture-influenced trust exists only within clans or kinship communities and unfolds outwardly step by step. According to the study by Fukuyama, all societies, deeply influenced by Confucianism, such as Korea and many regions of Southern Europe and Latin America, advocate "familialism" with the emphasis on rituals, participation, and kinship ties (Fukuyama 2015). Such a trust pattern classifying different levels of trust based on kinship widely exists in human society.

The acquaintance society takes an acquaintance community as a basic trust group bonded by the mechanism of "reciprocity". It follows the principle of extending different treatment to insiders and outsiders, and the trust is gradually extended outwardly. Firstly, acquaintance is the prerequisite for social trust in such a society, so the familiarity between acquaintances helps form a relatively symmetrical pattern of information mastery while the uncertainty caused by the asymmetry of information can be avoided naturally. Not only does such reliable interaction maintain a

---

[1] The term "Differential Pattern", coined by Fei Xiaotong, a renowned Chinese sociologist, is used to describe the concept of traditional Chinese interpersonal relationships. According to Fei, the pattern of traditional Chinese relationships "is not a bundle of clearly tied firewood, but like the ripples of a stone thrown into water waving in circles…centered on "self", which become farther and thinner in the process." (Fei Xiaotong. *Earthbound China*. People's Publishing House, 2008, pp. 28–30).

temporal and spatial continuum, but the trust between them is guaranteed by emotions and morals, which can both reinforce trust behavior among its members and provide certain degree of restraint on the breach of trust (Bo 2019). Secondly, the trust between people in an acquaintance society is sustained by human affection, which can be maintained on a set of "reciprocal" relationships. Such a society oriented at interpersonal relationship is more emotional than utilitarian, which maintains the productive order of the acquaintance society. Individuals are trained by the mechanism of "reciprocity", while they also enjoy rights and supervise interpersonal relationships, making the acquaintance society a network of micro-rights relations. People are integrated into a mechanism of associated interests and responsibilities to form a close community with few internal conflicts and high external solidarity (Bofeng 2011). Lastly, extending different treatment to insiders and outsiders is the basic principle in the acquaintance society. In a self-sufficient small-scale farmer economy composed of acquaintances, each family is self-centered and extends from the inside out. Different degrees of ethical norms and moral imperatives are developed according to different affinities between individuals, which goes that honesty and trust are required among people closely related, but not among strangers.

With transforming social structure and economic system, the acquaintance-based mechanism relying on cultural customs, moral standards, and interpersonal kinship has stepped into anomie to some degree. On the one hand, socio-economic activities carried out in acquaintance networks often end up in "*Sha Shu*" (taking advantage of acquaintances). Market economy has made economic achievement the criterion for value assessment and social stratification, and interest-chasing rational men may defy the mechanism based on reciprocity any time. Therefore, the negative functions, limits, and risks in modern society of relationship capital are increasingly visible. On the other hand, frequent social mobility encourages acquaintanceships. Such interest-linked relationship is weak in its emotional basis at the outset, which aggravates acquaintance interactions (Guangfei 2004). Moreover, "only when individuals were gregarious and cooperative did they begin to evolve into a civilized species. Only then can knowledge and technology be transferred, updated and extended" (Guanjun 2019). While family-based traditional culture leads to a high level of trust and dependence within the family, the ethical radius of trust is usually too small, spreading to "friends" who are nominally acquaintances at most. Confining trust within families will not only hinder the spread of social trust, but sustainable social prosperity.

3.2.2 Trust System in a Stranger Society

With the refinement of labor division, rapid transit system and the intergenerational occupational change in modern societies, the stranger relationship driven by production and exchange came to replace the acquaintanceship based on blood and geography and be the fundamental part of social relations. Anthony Giddens, a British sociologist, describes the feature of this transformation with the concept of "delocalization". He argues that delocalization mechanism, which relies on trust, allows social action to be "extracted" from territorialized sceneries and reorganize social relations across vast time and space distances (Giddens 2011). In an open society, interactions between people are more frequent while the sense of strangeness is cemented. The difference between strangers and acquaintances is no longer determined by the frequency and number of interactions, but by the openness of society. In fact, industrialized production relations impact acquaintanceships, because people familiar with each other will gradually alienate with reduced connections and common actions, and the acquaintanceships will be marginalized in social interactions. "Globalization entangles presence and absence, allowing distant social events and social relations to intertwine with local scenes" (Giddens 1998). The heterogeneity of social structure and framework dictates the gradual differentiation and pluralizing of social order, social norms and common values, which poses new challenges to the trust between strangers.

"In a rationalized market economy, while rituals and relationships still govern thought and behavior, more institutions should be given full play in a broader social scope" (Jianmin 2005). Institution-based trust is trust in institutions (including rules, regulations laws, and ordinances) that are commonly deemed as effective in social sphere, with which the trust in economic system, expert system, and legitimate political power can be secured. Therefore, it is more universal and binding beyond the scope of individuals and groups.[2] Firstly, institution-based trust adapts to the objective requirements of market economy, which is a combination of the rule of law and moral economy. Institution-based trust functions as a remedy for the inadequacy of ethics and morality in binding interpersonal relations and improves the stability and effectiveness of market economic activities. Secondly, institution-based trust unshackles the restrictions of blood, geography, and industry, and expands the scope of social trust in which it is possible for any individual, organization, and nation to build trust with others, so that it has wider application and adaptation. Finally, institution-based trust protects the interests of two or more parties in social interactions and reduces trust cost. The coercive, binding, and authoritative nature of institutions leads to less individual prediction of future behavior, thus reducing losses by trust risks (Jianmin 2005).

Contract-based trust is the core of institution-based trust, the safeguard mechanism of a stranger society, and a means of social order. Stability and harmony are the social ideals that people in all ages have been seeking, as an orderly society provides a settled production and living environment. Order and trust go hand in hand since the start of an age of civilization. As G. Simmel said, "Without trust, there is no way to construct a society, not even the most basic human relationships". There is naturally a basis for building trust between acquaintances, who develop direct trust based on morality and emotion. But such foundation is missing between strangers, and a bridge must be built for mutual trust. Such indirect trust relies on an intermediary called contracts. Contract-based trust focuses on issues instead of on persons, which avoids

---

[2]Chen Xin. *Cooperation in Social Dilemmas: The Power of Trust.* Science Press, 2019, p. 151.

affinity entanglement and monopoly, and discards the tedious process of "tuft-hunt" or "securing advantages through influence". A contract will be reached on the basis of rational calculation to maximize the benefits of both parties in the transaction and to establish a set of relationships based on rights and obligations. Based on such a contract, both parties will have expectations for each other with a belief that the other party will fulfill its obligations in future actions in accordance with the contract. This simplified trust-building process makes it possible to establish social trust and enables people to easily develop trust to cope with modern life and maintain modern social order.

3.2.3 Trust Crisis in a Network Society

The popularity of Internet in recent years has changed people's access to information and attitudes toward social communication. Meanwhile, public awareness and social trust have been under impact. The parallel existence of virtual cyberspace and physical world is increasingly evident in modern society, industrial organizations, and business formations. The cognitive and behavioral characteristics of netizens in online interactions tend to be influenced by their value orientation, personality traits, cultural backgrounds, and other factors in the real world. The essence of online interactions based on technological intermediaries is still "de facto communication" (Hua 2008) of individuals in the real world relying on Internet technology. With powerful science and technology, the network society is infusing communication with more diversity, dynamism, and universality. Information integration in the network era has facilitated the integration of different ideologies as a new type of modern team consciousness of "harmony in diversity". While facing real-time computers and media, people are receptive to a full range of cultural transmissions, leaving behind the redundancy and complexity of conceptual constraints of old customs, and having higher tolerance and recognition of anything foreign. An increasing number of people are perceiving from global cooperation and development beyond ethnicity and nationality; therefore, the "internet relationship" has become a new term for social relations apart from blood relations, geographical relations, and industrial relations.

Lawrence M. Friedman, an American sociologist stated in his *Law in America: A Short History* that "Our health, lives and wealth in the contemporary world are at the mercy of people whom we have never met and will never meet". Firstly, the network society is just a typical stranger society, in which the "trust of strangers" depends more on individual judgement, competence, knowledge, and morality, and one-way trust solely out of assumptions and self-assertion will undoubtedly take a huge risk. Secondly, the gradual weakening of moral binding in a network society is the main internal factor that triggers trust crisis. Away from the ethical situation of trust in daily life, the subjects of communication in a network are subject to the lack of moral authority, which leads to vague, undisciplined, and casual moral standards and responsibilities of the members inside, and a strong moral relativism permeates the mental state and moral life of the whole society. Finally, the trust crisis in the network society also has a bearing on the reliability of online information. Internet trust mirrors people's tendentious faith and behavioral choices in facing with uncertainties in the interaction among people and systems (technological platforms) in the digital environment, i.e., a "confident expectation" (Jianbin 2010). People's overall perception of the reliability of online information and the security of online activities, as well as the concern over the compliance of particular systems online with laws and ethics, will bring about potential and unpredictable effects and uncertainty to the network.

Under the influence of social risks, "trust becomes a crucial strategy in order to cope with uncertain and uncontrollable future" (Sztompka 2005). While trust crisis has been devastating social and individual development, it reflects the inadaptability of traditional trust model in modern civilization on the one hand; it exposes the shortcomings of current society on the other, thus prompting us to review human existence and social development. As a main medium for information, the role of Internet in guiding public opinion has been prominent. Factors including inadequate network restraint mechanism and morality giving way to interests have frequently led to a series of misconducts or incidents. Trust crisis upon systems, experts and media in various aspects of the society to different degrees has aggravated the trust crisis in the network. For example, with the construction of a well-off society in an all-round way, China's public welfare and philanthropy is progressing swiftly. Meanwhile, it is facing problems and challenges. Public charity is often only at the tip of public tongue, which weakens the brand image and credibility of social welfare organizations, and even eroded the foundation of public trust in social organizations. It is a requirement of the historical development of mankind and also the impetus for the sustainable social progress to eliminate trust crisis and restore trust in real life and in the network, thereby building a society of trust.

3.3   The Consensus Mechanism of Blockchain

Consensus is significant for any institutions at any time and in any region because of its role in maintaining social operation and consolidating a society. Blockchain, as a consensus mechanism based on the common recognition and witness of every transaction by all participants, involves three types of consensus in different contexts, namely, machine consensus, market consensus, and governance consensus. These three, in their entirety, determine the critical characteristics of blockchain of security, scalability, and decentralization, (Yizhong 2019) and play a role in social integration[3] and market stabilization.

---

[3] Social integration refers to a process by which a society, through various means or media, combines various elements, parts and links of the social system into an organically coordinated and unified whole to enhance social cohesion and integrity.

Machine Consensus. The key to blockchain technology is the design of a consensus mechanism, which aims to enhance security, scalability, and performance efficiency and to lower the energy consumption of blockchain to make it possible for strangers in the digital world to reach credit consensus through contractual mechanism rather than intermediaries. Consensus mechanism is the core of the distributed system. "Good consensus mechanism improves the performance efficiency of the blockchain system, provides strong guarantee in security, supports functionally complex application scenarios, expands and extends blockchain technology" (Xuan et al. 2019). In P2P networks (peer-to-peer networks), nodes that do not trust each other ultimately achieve data consistency by following a pre-defined mechanism, which is known as machine consensus. Machine consensus allows associated machines to connect to work and function despite the failure of certain members. Blockchain uses different machine consensuses, having different effects on the overall performance of the system while ensuring consistency and validity. The technical level of machine consensus can be evaluated from four dimensions. First, security, i.e., whether it can avoid attacks such as secondary payments, private mining, and whether it has good fault tolerance. In achieving consistency in a blockchain system driven by financial transactions, the main security issue is how to detect and prevent the secondary payment behavior. Second, scalability, i.e., whether it supports network node scaling. Scalability is one of the key factors in blockchain design. According to different objects, scalability can be divided into two parts: the increase in the number of system members and the increase in the number of transactions to be confirmed. Third, performance efficiency, i.e., the time from when a transaction is agreed upon and recorded in the blockchain to when it is finally confirmed, which can also be understood as the number of confirmable transactions per second in the system. Fourth, resource consumption, i.e., the computational resources that the system consumes in reaching consensus, including the central processing unit (CPU) and memory. The consensus mechanism on blockchain leverages computing resources or network communication resources to reach consensus (Xuan and Yamin 2017).

Market Consensus. A market society based on commodity exchange is a heterogeneous society that supports the pursuit of special interests by independent individuals, while requiring social harmony premised on the coexistence of diverse interests and values. For value consensus in modern society, understanding must be achieved via the establishment of agreements with common norms in interactions, and thus reaching recognition and consensus. A society cannot exist without consensus. At the most basic level, consensus is a foundation that keeps teams of people out of conflict and drives them to make decisions together. In the era of big data, it has become a consensus to adhere to one security standard, one benchmark, to be impartial to various stakeholders in cyberspace, and to balance the interests of all participants. Firstly, market consensus is reflected in the equilibrium price formed in market transactions. Take bitcoin as an example, the fundamental reason why it is increasingly accepted is that blockchain technology provides a widely recognized and accepted consensus mechanism that gives it the function and utility of a fiat currency in a given market environment. "Whether bitcoin can be used as a currency or not is based on whether there is a monetary identity between parties, which determines whether blockchain can be used in a wider range of applications than bitcoin and exert greater technical advantages and institutional value" (Lei 2018). Secondly, the blockchain is both a "consensus system" for multiple parties and a benign gaming mechanism. Along with the progress of technologies, the consensus mechanism has gradually developed from an abstract concept into an important support for distributed ledger technology, which underlines that all or most members of the network shall agree on information about certain transaction or on certain piece of data. Consensus among participants is at the heart of the blockchain. In the absence of a central authority, participants must make an agreement on rules and applications, and agree to apply those rules to transactions. Those who follow the rules will benefit and those who break the rules will be evicted. Finally, the security and irreversibility of distributed data storage records must be ensured in the shortest time in order to achieve a decentralized system that is secure, reliable, and immutable. In a mistrustful market, a necessary and sufficient condition for the unanimity of nodes is that each node aims to maximize its own interests, follow the pre-set rules of the agreement willingly and honestly, determine the truthfulness of each record, and ultimately write down the real record into blockchain. It is fair to say that the basis protocol of market mechanism stimulates each node's activity to create wealth through price and competition, and makes it possible the large scale of collaboration among mutual-distrusted nodes, thus unlocking the enormous potential of sharing economy and collaborative governance.

Governance Consensus. For a society void of the rule of law in the real sense, an ideological movement targeting at consensus is a major impetus for social changes. Simply put, for proper society functions, we need to reach "consensus on facts". Governance consensus, one of the three elements of a blockchain consensus system, means that all the members develop and agree on a most advantageous decision in group governance. There are four key elements, namely, different interest groups, a certain governance structure and rules of procedure, reconcilement and compromise between conflicting interests or opinions, and group decisions that are generally binding members. Governance consensus involves subjective value judgments of individuals and deals with subjective multi-value consensus. Participants are reconciled through intergroup coordination and collaboration to converge to a single opinion, and failure to do so would lead to a failed governance consensus (Zhong and Chuanwei 2018). Institutions responsible for these basic facts already exist in the developed world, but they are now being widely criticized. Tomicah Thielemen, president of the Global Blockchain Business Council, noted that blockchain is subject to erosion-resistance and a new landscape that facilitates agreement on core facts while ensuring the confidentiality of privacy-related facts. Blockchain allows a group of people to agree on a variety of facts without relying on the arbitration of a centralized entity. The history of human civilization comes not from so-called absolute facts, but from a much more powerful concept–consensus, an agreement inside the society which allows people to discard doubt, build trust, and interact in a collaborative way. "Technology and law are substitutable in national and social governance. If technological solutions in a particular social scenario cost less than legal solutions, technological tools may then replace legal forms as the primary means to generate order" (Ge 2018). Blockchain, as an underlying technological framework and consensus mechanism with universal application, may bring about profound changes in finance, economy, technology, and even politics.

3.4 Trust is De-Trust

Under a new wave of technological and industrial revolutions, blockchain is emerging globally as one of the key technological headwinds. Decentralization, as a fundamental feature of blockchain systems, enables the exchange between nodes to follow a fixed algorithm without the need for trust. The establishment of a decentralized transaction mechanism that does not rely on the trust from third party and cannot be manipulated has become a major feature of blockchain in the value-interconnected system.

3.4.1 Centralization and Decentralization

Mankind initially expected the Internet to be a utopian which could bring about an equal environment. Friedman, a columnist of *New York Times* argued that the world is delayering (Friedman 2008). From the era of personal computer (PC) to the era of mobile devices, business and government management transitioned from a pyramid organizational structure to a delayered one, which is an idea of decentralization. The Internet has led us down the path of decentralization even before the birth of blockchain. A decentralized network similar to BitTorient appeared as early as in 2000, and cryptographers, mathematicians, and software engineers have also been working for nearly three decades to advance protocols, thus achieving a stronger privacy and trustworthiness for e-cash, voting, file transference and storage, as well as other types of systems. However, the Internet has never been able to address information ownership, and the innovative invention of blockchain is precisely the missing piece of the clue, which connects the researches on decentralization and cryptography, and then "gives it a whip" to drive the industry to leapfrog.

Blockchain is not just a decentralized ledger for storing processes and results. It is also a multi-purpose decentralized platform after process reconstruction, which will bring about a significant reduction in "transaction cost". Such cost refers to the economic cost of bureaucratic and managerial interactions for clarifying the details of a legal contract, ensuring the credibility of the counterparty, and documenting outcomes. It also explains why a decentralized model of interaction is preferred over interaction with aggregated centralized large firms. We may find out that the economic pattern in twenty-first century is more like that in eighteenth century, which replaces the mode of a single firm selling insurance products with mutual guarantees, the centralized third party clearing financial transactions and payments by peer-to-peer transactions, and even evaluates trust and reputation, conducts quality control, and achieves property rights tracking with a more decentralized approach. Meanwhile,

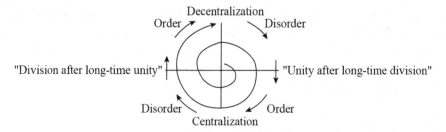

Fig. 3.1 The division and union of the world system. *Note* The world system also follows the logic that "The empire, long divided, must unite; long united, must divide". It moves from one center to another, and then develops into decentralization, or multi-centeredness. *Source* Linstone H, Mitroff I. *The Challenge of the 21st Century*. New York: New York State University Press, 1994

the extreme efficiency of information technology in twenty-first century makes it possible to achieve such a social form at a very low cost (Hang 2016).

The exploration of blockchain is not simply aimed at decentralization. It might be polycentric or weakly centered, with the end result more likely to be polycentric in a bid to avoid the loss of control caused by too much discursive power possessed by a few subjects (Chang Jia et al. 2016). "Decentralization" implies a kind of "distribution", and blockchain brings new technologies to the consistency in distributed systems and mutual trust, mutual recognition, and interconnection worldwide. It is the law of history that "The empire, long divided, must unite; long united, must divide", and the philosophy of "going too far is as bad as not going far enough" should be avoided in world development (Fig. 3.1). Centralization lacks transparency and data credibility, while decentralization produces higher energy consumption and cost. In blockchain applications, it is logical to choose different types of chains, different consensus mechanisms and other technological solutions according to different degree of decentralization needed to avoid excessive pursuit of decentralization on a certain extent (Table 3.1). In such a context, a more harmonious and stable global order will be achieved.

3.4.2 De-Trust of Blockchain

Replacing intermediaries with technology as a "trust machine" is an enduring pursuit of human society, for which a new development platform has emerged, thanks to the birth of blockchain. Mankind is moving from a physical society that has lasted for millennia to an emerging society constructed by virtual numbers (Sanderson 2015). In traditional societies, trust has been built primarily through credit endorsements by third-party credit service agencies, but such private and centralized technical architecture cannot fundamentally achieve mutual trust and value transmission from online to offline in the digital age. In contrast, blockchain, with a decentralized database architecture, can accomplish trust endorsement of data interactions and

Table 3.1 Classification of Blockchains

Project	Public chains	Union chains	Private chains
Participants	All	Union members (e.g., public security related department)	Within individuals or organizations (e.g., the confidential unit of public security organizations)
Access rights	Anonymous	Registration and permission required	Registration and permission required
Degree of centralization	Decentralization	Polycentric	Partially centered/weakly centered
Consensus mechanisms	POW (proof of workload), POS (proof of stake), DPOS (delegated proof of stake)	PBFT (practical Byzantine fault-tolerant algorithm), RAFT (consistency algorithm under distributed surround)	PBFT (practical Byzantine fault-tolerant algorithm), RAFT (consistency algorithm under distributed surround)
Incentives	Needed	Adjustable	Not needed
Applications	Bitcoin, Ethereum	R3 Bank Interbank Union	Ark Private Chain
Features	Open, transparent	Efficient	Secure, traceable

Source Zeng Ziming, Wan Pingyu. Big Data Resource Management System Based on Sovereign Blockchain Network in the Field of Public Safety. *Information Studies: Theory and Practice*, 2019, No. 8

build trust between any participating nodes in the network relying on their recognition of consensus, thus building algorithmic trust. As described in "The Trust Machine", a cover article in 2015 *The Economist*, blockchain is a trust-making machine, and plays its role in any area where trust is needed.

Blockchain improves production relationships within existing technologies. Seen from history, people, or institutions and organizations made up of people, are always most untrustworthy in trust systems. History often ends up proving that those who break rules are rule makers. In *The Three Body Problem*, a well-read sci-fi, there is a Mock Negotiation[4] in which the "suspicion chain",[5] once initiated, will inevitably

---

[4]Dry Running: Both A and B want to achieve peaceful coexistence in the community, but A will not feel ease even if A believes B is well-intentioned, because the well-intentioned person cannot think of others as well-intentioned in advance. In other words, A does not know what B thinks of him, or whether B deems A as well-intentioned. Furthermore, even if A knows B thinks A as well-intentioned, he will never know how B thinks of him. The same is for B. "Pretty roundabout isn't it? This is only the third level, and such logic can go on and on endlessly." This means that as long as there is suspicion of others, the suspicion chain will kick off and can never be closed.

[5]The most important characteristic of the suspicion chain has nothing to do with the social form or moral orientation of the civilization. Each civilization shall be taken as a point at either end of the chain and that all civilizations become one and the same, regardless of whether they are well-intentioned or malicious within themselves, when they enter the network formed by the suspicion chain.

lead to "eternal life of death". The "suspicion chain", in effect, is a scenario full of mutual distrust, or more precisely, the logical consequence of inability to establish original trust, which means each individual (nation or civilization) is reluctant to make further "overdrafts" beyond its known information and rational evidence. The logical individuals in the chain remain what Hobbes calls a pre-political "state of nature" and have to tear each other apart like wolves. Nowadays, "tearing" has become an evident scene among all the people and permeating every corner, in the real world or on the Internet. Ordinary people, celebrities, and even presidents of great powers are all tearing each other whether inside families, on courts, or on the social media of Weibo or Twitter (Guanjun 2019). It can be concluded out of the wave of technological development from the industrial revolution to the Internet revolution that the great leap of productivity is usually achieved by replacing human being, which is considered the most unreliable, vulnerable, and the least efficient link. Based on codes and passes, the blockchain value ecosystem delivers full information credibility and a convenient mechanism for the flow of benefits, in which members can exchange and collaborate better with immediate interest feedback, thus maximizing incentives for ecosystem participants and involving more people into the collaborative state quickly.

"De-trust" does not mean that trust is not needed any more, but that trust no longer needs to be provided by a traditional centralized third-party authority. What the blockchain enables is a shift in trust from people and institutions in collaboration to the blockchain as a consensus machine. Based on this, it can be concluded that the essence of the blockchain mechanism is not "de-trust", but recreating trust. The core of such mechanism is to achieve consensus, common recognition, and common management of all operations through consensus. The consensus mechanism, based on cryptography and code, is aimed at reducing communication cost to the minimum through the Internet and the common preferences of participants. It's optimal to regard blockchain technology as a tool that society uses to build larger-scale trust, create social capital, and bring about the common story needed for a better world, but not simply a tool to replace trust.

3.4.3   Blockchain-Based Credit Society

Based on the blockchain, mankind is establishing a complete set of governance mechanisms for credit Internet. Blockchain technology has wiped out information asymmetry and subverted trust and credit in the traditional sense by building trading rules based on trust in technology and low-cost credit mechanisms. Big data Internet companies did not generate much-needed decentralized credit resources for the global credit market, with centralized data. But with the open-sourced and transparent blockchain, participants can verify the authenticity of the ledger history, which helps to circumvent the incidents on P2P lending platforms, such as escaping with money and fraud. Moreover, the process by which blockchain transactions are confirmed is the process of clearing, settlement and auditing, which can enhance efficiency. The greatest appeal of blockchain lies not in its power to change the rules by which the world works, but in its potential to extend the freedom of individuals, that is, the individuals can retain the incentive to commit fraud, or even commit fraud, yet the last firewall, the blockchain mechanism, can eliminate the possibility of harm caused by fraud and breach of contract in any nodes.

Blockchain holds great promise for addressing the current social pain of scarce credibility and can be widely and deeply applied in global market convergence, intellectual property protection, property micro-notarization, IoT finance, social welfare industry, and many others. The data stored on the blockchain is highly reliable and tamper-resistant, naturally suitable for use in social welfare. The reasons are as follows. First, the relevant information in the process, such as donation projects, collection details, fund flow, feedback from recipients, can be stored in the blockchain and conditionally be made public in the premise of protecting the privacy of participants and abiding by relevant laws and regulations, which can facilitate public and social supervision and contribute to the sound development of public welfare (Maolu and Jingyi 2018). Second, the decentralized blockchain helps reduce the costs and time in the acceptance of donations, which can improve efficiency. Donors can make peer-to-peer donations to charities by purchasing cryptocurrencies issued on the blockchain, which is convenient and safe, and expands channels and increases donation. Third, blockchain technology maintains a consensus mechanism that keeps the consistency of data across blocks with the open nature and instant information sharing capabilities, which can disburden information sharing and the cost of information system operation, clarify the respective permissions of chain members, and avoid redundant information to solve problems from the root such as duplicate reporting of information (Han 2018).

"Blockchain is the fourth milestone in the history of human credit evolution, after the credit of blood relatives, of precious metals, and of central bank notes" (Ruofei 2016). The most possible change that blockchain is about to bring to humanity is a new credit society, for blockchain technology is a "credit technology" and a credit infrastructure in the digital world and virtual society. The decentralized, transparent, and open blockchain establishes "credit" through whole network of bookkeeping and P2P collaboration. Instead of "digital currency", blockchain takes as its core an ecosystem building "credit" in an uncertain environment, which to a certain extent mirrors the Internet thinking and the concept of "everyone-to-everyone society" (Jiankun 2018). Rather than merely hoping that the adversaries we interact with behave well, blockchain systems embed the property of trust inside, so the system can function well even if many of the participants are misbehaving. Blockchain activates the credit machine for us, allowing governments, companies, and other institutions and individuals to be present as equal nodes in a distributed network. Each of them manages its own identity and credit, shares a non-modifiable transaction ledger, and helps reshape the mechanism, transform the process, enhance trust, and improve efficiency during the governance.

3.5 Digital Trust Model

Bernard Barber, an American sociologist of science, noted, "Although trust is only a tool of social control, it is a ubiquitous and important one in all social systems". The reason lies in that trust is the bridge connecting individual citizens to the community. History has taught us that dreams and institutional designs cannot reshape trust between groups of people, and that trust is to be based on reliable technologies. Blockchain is just targeted at trust mechanism in human society at this point. Digital trust is a picture of de-trust that caters to the needs of the digital age, an advanced form of personality trust and institutional trust, and an approach for value transmission and mutual trust. The digital trust, personality trust, and institutional trust are co-existent rather than substitutive. However, description alone is insufficient to define the nature of digital trust, which may easily lead to a single, conceptualized understanding. Therefore, it is necessary to build a digital trust model in terms of an overall picture, internal mechanism, and operational process, thus to understand digital trust in a scientific and systematic way.

3.6 Model Theory and Trust Model

Model and Modeling are Two Core Concepts of Model Theory. Model is primarily the abstraction of objective issues and laws, and finally it is applied in practice. Magnus Hestenes in his model theory pointed out that the model is a conceptual representation of the real object, a substitute for the object (Hestenes 1992). The Israeli scholar Heusler and the American scholar Smit emphasized "Model is not a true depiction of the system, but a set of hypotheses made to explain certain aspects that are objective reality, and that models are only temporary workarounds for visualizing and explaining objective reality" (Wenqing 2011). According to Chinese scholars Jiang Xuping and Yao Aiqun, a model is a formalized expression after the abstraction of a practical problem or an objective object or law. There is a wide variety of models, with common categories including mathematical models, conceptual models, structural models, systems models, procedural models, management models, analytical models, method models, logical models, and data models. An appropriate model is chosen on modeling objectives, variables, and relationships. Among the many types of models, a mathematical model is usually a set of mathematical expressions that reflect the laws of operation and trends of change of an objective. Conceptual models are represented, for example, in the form of diagrams expressing concepts. Structural models mainly reflect the structural characteristics and causal relationships of the system, in which graph models are an effective way to study systems, especially complex systems, and are often used to describe the relationship between nature and objects in human society. Modeling is to describe complex systems, illustrate the relations between objects and enhance people's understanding of patterns. It can help analyze and describe complex systems in a detailed fashion so as to upgrade people's control over details, and to see through appearances to perceive the essence so as to systematically understand the connections between objective things and the effects they have. With respect to the steps of modeling, rather than the five steps of model selection, construction, validation, analysis, and extension espoused by Harlan in his book *Diagrammatic Modeling and Concepts of the Real World: The Concept of Newtonian Mechanic*, the seven steps of preparation, hypothesis, composition, solving, analysis, testing, and application are more often followed nowadays.

Traditional Trust Model: a Trust Model on Personality Trust and Institutional Trust. Based on a number of factors that influence the generation of trust, Meyer, Davis, and Schumann attributed the trust generation to trustor's internal predisposition and trustee's trustworthiness such as perceived competence, goodwill, and integrity, upon which they construct a lean and concise model of interpersonal trust. They argue that trust is an interpersonal psychological interaction, and that only by considering both sides of a relationship can interpersonal trust be accurately illustrated and the insufficiency of previous research be filled. Roper and Helms introduced the concept of motivational attribution[6] and construct a trust component theory model,[7] which discusses the change of trust and motivational attribution in intimate relationships. They thought the three components of predictability, reliability, and belief are not mutually exclusive, but represented different types of trust at different stages of intimacy. It is just that at each stage of the relationship, the proportion of each component varies, and there is necessarily one component that is in the absolute leading position. Such differences also impact the motivational attributions of the participants and ultimately distinguish the stability and the emotional connections of intimate relationship. In his analysis of reasons for trust, Piotr Sztompka argued that the culture of trust is an independent, given, and explanatory variable.

---

[6]Attribution Theory: In order to effectively control and adapt to the environment in daily social interactions, people tend to consciously or unconsciously make certain explanations for various social behaviors in the surrounding environment, i.e., cognitive subjects infer other unknown characteristics based on some specific personality traits or some behavioral characteristics of others in the process of cognition, in order to seek a causal relationship between various characteristics.

[7]Generally speaking, trust arises on past experiences and matures as the relationship gets closer. In an intimate relationship, the individual develops some specific motivational attributions to the other party with growing trust and begins to be willing to take the risk of this attribution failure. In this sense, trust is more frequently defined as a sense of security and confidence that arises from intimate relationships. Based on these characteristics, Roper and Helms, after combining the various academic studies on trust, propose that trust consists of predictability, reliability, and belief. Predictability is mainly based on the experience of social interaction between the two parties, and on the consistency, continuity and stability of individual behavior. Reliability means an individual stops being entangled in the particular behavior of another person but moves to making a holistic judgment about the other person's motivations and personality traits. This judgment is reflected when the individual begins to think about whether the other party in the relationship is trustworthy and reliable, and whether it enables the individual to produce a sense of security, etc. Belief embodies the highest level of trust, mainly reflected in the emotional certainty and reassurance that the individual holds upon others. In a risky future society without concrete evidence of support the action, the individual still holds confidence in others and believes the other party in the relationship will meet his needs and base actions on his welfare and better development.

He set normative consistency, stability of social order, transparency of social organization, familiarity of social environment, responsibility of people and institutions as the five macro-social contexts, and social emotion and collective capital, which are two factors in historical dimension and the individual contributions of actors, as basic elements, and construct the social generation theory model of trust culture.[8] Based on the research fruits related to network trust, Lu Yaobin, Zhou Tao, and other scholars proposed that according to research contents, network trust model can be divided into initial trust model, system-based trust model, virtual community trust model, B2B (business-to-business) website trust model, online store trust model, etc. Online trust has received extensive attention in both social and natural sciences. Scholars in different disciplines have explored it theoretically and empirically from various perspectives and drawn different and even contradictory conclusions. Therefore, the fundamental problem of network trust remains as open as offline trust. As Georg Simmel stated, "Society itself would become a mess without general trust between people, since there is never a relationship which is not based on the certain knowledge of others". The same is true of collaboration and trust. Collaboration is the result of decision-making, so it is intuitive, while trust is an attitude that accompanies the decision-making process and is endogenous. It is fair to say that the existence of trust may be inferred between collaborators based on the collaborative behavior they exhibit, but this does not mean there is trust between them. As we move into the digital age, where more and more highly complex tasks rely on the collaboration of decentralized, independent, and contextually diverse participants and web service platform, traditional trust model can no longer keep up with the times.

Digital Trust Model: A Decentralized Trust Model on Blockchain. "Blockchain is a new type of decentralized protocol, on which the data cannot be changed or fabricated at will, thus providing a credit paradigm without the need for trust accumulation. Blockchain can be understood as a ledger where people simply need to join an open and transparent database and reach a credit consensus through peer-to-peer bookkeeping, data transfer, and authentication or smart contract without resorting to any more intermediaries" (Jia et al. 2016). Hyperledger, cross-link transmission, and Smart Contract is three key steps in digital trust (Fig. 3.2). The hyperledger endows data sufficient trustworthiness, so it is a fundamental link in the formation of digital trust and plays an essential role for data to enter cross-link transmission and smart contract. Cross-link transmission is a crucial step in digital trust, as it is an upgrade of traditional Internet shifting from information transmission to value transmission. Smart Contract, a core link of digital trust, does not need third-party trust endorsements, so it achieves decentralized trust. We refer to the specific implementation of digital trust as digital trust model, expressed by the following

---

[8] Social Generation Theory means that traces of previous events accumulate in institutions, rules, symbols, beliefs, and the psyche of social actors. Shared experiences give birth to shared structures, cultures and psychological patterns, which in turn provide conditions for future action. The theory argues that human action is the driving force behind social processes and that actors' actions are both constrained by social structures and reproducing new structural conditions. And the reproduced structures will in turn become the initial conditions for future practice. This process is infinitely circular and open to all possibilities.

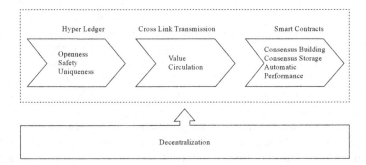

Fig. 3.2   The process of how digital trust achieves value transmission and mutual trust

formula:

$$T = H(L, C)。$$

in which T stands for digital trust model, H stands for hyperledger, L stands for cross-link transmission, and C stands for smart contract.

3.7   Hyperledger: From Traditional Bookkeeping to Everyone-To-Everyone Bookkeeping

All interactions in human society are based on certain degree of trust; however, it is slow to build trust, difficult to maintain it, and quite easy to destroy it. Hyperledger aims at establishing an open cross-industry standard as well as an open source code development library, allowing companies to create customized solutions to distributed ledgers to facilitate the use of blockchain in industries. Decentralized, tamper-resistant, and permanently traceable, it helps form a trustworthy digital ecosystem via distributed bookkeeping and free notarization across the network and lays a solid foundation for digital trust.

Distributed Storage Showing the Coming Future. Not a single technology, blockchain is a product of the integration of technologies from multiple fields, including cryptography, mathematics, economics, and cyber-science. Such integration creates a new decentralized data recording and storage system, and timestamps the blocks of stored data so as to create a continuous, consecutive storage structure of trusted data records with the ultimate goal of creating a highly trusted data system, referred to as a distributed database that guarantees a trustworthy system. In this system, only the system itself is trustworthy, so data recording, storage, and updating rules are designed to establish people's trust for the system. At present, digital technologies, represented by artificial intelligence and blockchain, are emerging and swiftly penetrating all economic and social sectors. The data-centric digital trans-formation is accelerating, with new applications in all industries soaring from thou-sands and tens of thousands to millions in numbers, and data becoming increasingly massive, diverse, real time, and multi-cloud. Data storage has become the basic plat-form for industrial transformation, and distributed storage has become a trend, which will account for 40% in 2023 (Inspur, International Data Corporation 2019). Earth-shaking changes and far-reaching impacts it will bring to all areas of human society is immeasurable, which is likely to be far greater than any other technological revo-lution in history. It will be a milestone in human capability to construct a network that is truly trusted and a huge step forward toward global mutual trust and a digital trust society through decentralized technologies based on big data.

Hyperledger is Characterized by Openness, Safety, and Uniqueness. "Open-ness means the data stored in the ledger is completely open to all participants, because of the peer-to-peer network storage method of the blockchain. In a blockchain network, every node stores a copy of the blockchain, and the uniqueness of the ledger ensures the copy is precisely identical across nodes. Safety means the data on the ledger is stored through digital encryption, and only members who hold corre-sponding decrypted data (private key) can interpret it, while others can only see and verify the integrity and uniqueness of the data, but do not have access to the private key. Miners all over the world are constantly hash colliding, improving the hash value and making it much more difficult to avoid any cyberattack. Uniqueness means the data stored is immutable. It includes both spatial uniqueness which means each node has only one identical piece of data, and temporal uniqueness, meaning the historical data cannot be changed. Uniqueness also means there is only one unique chain in the operation, because if there are different chains, a bifurcation will come into being and bring about copies in two different spatial dimensions of blockchain, which is precisely what to be avoided by the consensus rules" (Jingdiwangtian et al. 2016). Under the blockchain protocol, no one can tamper with historical data, as it is openly shared in the blockchain. A series of data governance challenges can be effectively addressed, e.g., the data cannot be used, is not convenient or is difficult to be used, and cannot be used effectively. It will be extremely difficult to tell lies, and trust will become stronger.

3.8 Cross-Link Transmission: From Information Transmission to Value Transmission

Human is in the midst of a historical migration from the physical to the virtual world, and the gradual transfer of human wealth to the Internet is also an established fact. The essence of the Internet is interconnectivity and the essence of data is interoperability. Cross-link, simply put, means achieving the connectivity between different chains so as to transmit value from one to another. Unlike the information flow on the Internet, cross-link is not a simple information transmission, but rather a free circulation of value.

Blockchain Technology Accelerates Data Capitalization. "After nearly half a century of informationization, the amount of data has exploded, data processing capacity has increased rapidly, and the accumulation and exchange, analysis and application of massive amount of data heightened production efficiency. Data has now become an independent factor of production. Meanwhile, data elements are featured by economic characteristics that are different from material factors such as capital, labor and technology. Firstly, it's featured by immediacy, which means data generation is online real time, and its processing is fast and in steps with economic development; secondly, the provider of data elements won't lose the right to use date after its transference, for data elements can be shared; thirdly, data has an increasing marginal productivity, that is, rather than being consumed, new data will be generated during the use" (Yongsheng 2015). Data that can generally be replicated in bulk is difficult to be capitalized, while blockchain can put a special identity stamp on the data to rebuild its attribute, which can help exploit the value of data in the business world, making it unique and turning it into an asset of value. Blockchain, through the inclusiveness of data, is crossing the inherent boundaries between nations, governments, organizations, and the public, and maximizing our ability to create a "shared world", i.e., a world that we can only share no matter what circumstances we are in. This will be within reach once data capitalization systems are established through blockchain technology.

Blockchain is the Key Element in Value Transmission. If we say "Byzantine General Problem" reveals the difficulty in information communication and coordination between scattered individual nodes, blockchain is then the clearest, most powerful, and most realistic way to solve this difficulty by far. Blockchain can be seen as a system of technologies to a more reliable Internet system that basically wipes out fraud and rent-seeking in value exchange and transmission. "It offers another possibility of direct peer-to-peer interaction, just as what we did in primitive societies—sincere face-to-face communication, without intermediaries, sometimes even without mutual trust, only limited understanding, and coming and going in haste, leaving everything else to the bottom. Such a system makes transactions, indeed all interactions, easier and more efficient. From each single point, this is only an individual transaction, but from a higher dimension, it makes large-scale, non-intermediary and collaborative interactions possible. There is no longer a need for the centralized intelligent treatment of powerful, huge intermediaries, for intelligence is hidden at the bottom, in chains and everywhere. It is not behemoth-like, but it is omnipresent, as it will eventually reshape our business, culture, and future society" (Mingxing et al. 2016).

Cross-Link Technology Extends the Boundaries of Blockchain Application Ecosystem. "Blockchain technology exerts revolutionary impact. Fundamentally, its impact lies in its signal to reorganize the relationship between the virtual world and the physical world" (Naiji 2017). Cross-link technology is significant for changing the isolation of blockchain and enabling the flow of values between different chains. With cross-link technology, future blockchain will become an infrastructure like water, electricity, and roads today where all values can circulate freely. More importantly, cross-link technology unlocks the value potential between different chains.

Through the cross-link module, data exchange, value growth, scenario application, and interoperability between chains can be achieved simply and efficiently, finally creating a shared value ecosystem to form a value-added system.

3.9   Smart Contract: From Choice Trust to Machine Trust

Trust plays a unique and positive role in social integration and cooperation. From conventional transactional contracts to blockchain Smart Contract, social trust pervades both the physical and virtual worlds with the aim of preventing and mitigating risks and building social consensus in information asymmetry. Through blockchain Smart Contract technology, people's objects of trust in the social system will be transferred from people and institutions to the consensus machine of blockchain, and culture, institutions, or third-party intermediaries are no longer necessary. It can be argued that new technologies such as blockchain Smart Contract will not only enrich the theory of social consensus, but create an entirely new system of social cooperation, where trust and social cooperation will be given new meanings in the digital society.

Smart Contract and Conventional Contract. There are similarities between Smart Contract and conventional contract, such as the need to clarify the rights and obligations of participants, and the penalties for non-compliance. Meanwhile, there are also critical differences (Table 3.2). In terms of automation, Smart Contract can automatically determine the trigger condition and choose the next transaction, whereas conventional contracts require people to make judgement, so they are inferior to Smart Contract in terms of accuracy and timeliness of condition judgment. From subjective and objective perspectives, Smart Contract is suitable for the scenarios of objective requests, while traditional contracts are suitable for those of subjective

Table 3.2 Smart contract versus conventional contracts

Comparison dimension	Smart contract	Conventional contracts
Automation	Auto-judge trigger conditions	Manual-judge trigger conditions
Subjective and objective	Suitable for objective requests	Suitable for subjective requests
Cost	Low cost	High cost
Execution time	Pre-determined, preventive execution mode	Post-execution mode, determining the pattern of rewards and punishments according to the state
Default penalty	Rely on collateral, margin, digital property, and other collateral assets with digital properties	Rely heavily on penalties and legal means can be used to enforce rights in the event of a breach of contract
Scope of application	Globally	Limited by jurisdictions

requests. The agreements, collateral, and penalties in Smart Contract need to be clarified in advance, but the subjective judgment indicators are difficult to be incorporated into the contract automaton for judgment, thus it is difficult to guide contractual transactions. In terms of cost, the execution cost of Smart Contract is lower than that of conventional contracts. The execution rights and obligations of Smart Contract are written into computer programs for automatic execution, so they enjoy the advantage of low-cost in state judgment, rewards and punishments, assets disposal, etc. In terms of time, Smart Contract comply to a pre-determined and preventive execution mode, while conventional contracts follow a post-execution mode which determines the pattern of rewards and punishments according to the state. In terms of default penalty, Smart Contract rely on collateral, margin, digital property, and other collateral assets with digital properties, and in the event of a default, the participant's assets will suffer great loss. The penalties for default in conventional contracts rely heavily on penalties, and legal means can be used to enforce rights in the event of a breach of contract. In terms of scope of application, Smart Contract is applicable globally, whereas conventional contracts are limited by jurisdictions, with different laws, cultures, and other factors affecting its execution (Jia et al. 2016).

Blockchain Smart Contract: the key to consistency. "The consistency in distributed systems means giving a series of operations, achieving 'some' degree of synergy on processing results for multiple service nodes under the guarantee of an agreed protocol" (Baohua and Chang 2017). If distributed systems can be "consistent", they can present a perfect, scalable "virtual node" to the outside world, which is also the most desired goal of a distributed system. "Smart Contract, programs event-driven, stateful, multi-party-acknowledged, running on blockchains, are capable of automating the disposal of assets based on pre-defined conditions. Its biggest advantage is the use of program algorithms to replace human judgments in arbitration and contracts enforcement (Fig. 3.3). Essentially, a smart contract is a piece of programming, featured with tamper-resistance, transparent data and forever running".[9] The two major steps of a smart contract includes the participation of multiple users in the formulation of a smart contract, and the proliferation of the contract through the P2P network and its storage in the blockchain. Contract generation consists of four main steps: contract negotiation, specification development, content validation, and code acquisition. The parties involved will determine the contents of the contract, clarify rights and obligations through negotiation, after which the standard contract text can be determined and programmed, thus the standard code is produced after verification. It involves two key links: contract specification and contract validation. The former needs to be developed through negotiation among relevant experts and contract parties, and the latter is executed on a virtual machine based on abstract system model, which is important for the security of contract execution to ensure the consistency of negotiated contract text and contract code. The release of the contract, i.e., the signed contract is distributed to each node via P2P. Each node receives it and

---

[9]Information Center of the Ministry of Industry and Information Technology. China Blockchain Industry White Paper 2018. Ministry of Industry and Information Technology, 2018. https://www. miit.gov.cn/n1146290/n1146402/n1146445/c6180238/content.html.

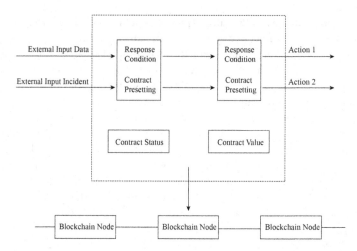

Fig. 3.3 Blockchain-based smart contract model. *Source* Ping An, China. A Panorama of Blockchain Industry. *Financial World.* 2020. https://istock.jrj.com.cn/article,yanbao,30771116. html

temporarily stores it in memory for consensus validation (Haiwu et al. 2018). Based on this, the contract terms are automatically enforced without relying on any third party when the trigger condition is met.

Consensus: a Guarantee for Machine Trust. Trust is the most significant value in cyberspace, and consensus is the greatest value of blockchain Smart Contract. "Social theorists argue that the reason for the orderly function of a large and complex modern society within a certain order is the development of a complex mechanism for social integration. The emergence of digital technologies, such as blockchain Smart Contract, has not only innovated the traditional idea of social trust, but opened up new possibilities for social cohesion and integration, i.e. the upgrade of social trust mechanisms to a certain extent. In social integration, people may not fully identify with the perspectives, values and perceptions of others, but we are able to inspire empathy when building trust between each other, so the society is able to get greatest common divisor" (Tianshu and Biao 2019). Blockchain Smart Contract, applying related technologies such as mathematics and cryptography, creates social trust and promotes social cooperation. The changes it brings about are unprecedented in terms of speed and scale, and it is critical for exercising full social and economic potential of digital technologies, reducing risks and avoiding unintended consequences. It goes without saying that blockchain Smart Contract exert far-reaching implications for social consensus and social integration practices.

3.10 Digital Currency and Digital Identity

"If the progress of human civilization is a spaceship, then high technology is the engine of its booster, which consists of five stages of propulsion, namely, digitalization, networking, miniaturization, simulation and an ever-powerful force" (Scott-Morgan 2017). The world today is still witnessing the coexistence of changes and constancy. In facing the changes unseen in a century, the questions of what is changing, in what direction it is changing, and what it will become are still uncertain and even unpredictable. But what is certain is that the power of digital identity and digital currency is changing our lives, and the world. Digital currencies will trigger sweeping changes across economy, and digital identities will reshape the mechanisms of social stratification.

3.11 Digital Currency

Currency is the physical evidence of a nation's creditworthiness, a representation of its overall strength, and a major guard of national sovereignty. Legal digital currencies must be guaranteed by national sovereignty. "Beginning with Bitcoin, digital currencies swept the world in 2009. If we see the progress of digital currencies as a game, Bitcoin is merely the button to start it" (Mingxing et al. 2016). In 2019, with the advent of the Libra,[10] we are once again witnessing the possibility of disrupting the world's monetary and financial systems. However, existing digital currencies, both at the technical level and the legal and regulatory level, still face numerous challenges and a series of problems with urgent need to be addressed. These challenges and problems can precisely help us think about and define sovereignty digital currencies.

3.11.1 Hazards of Digital Currencies

Currency is one of the greatest inventions in the history of mankind. According to Marxist political economy, currency is a commodity capable of serving as a fixed general equivalent. Modern monetary theory holds that currency is a contract between the owner and the market regarding exchange rights. With evolving commodity

---

[10]Facebook's Libra, a cryptographic digital currency that does not seek stability against the US dollar, but rather relative stability in real purchasing power, is initially collateralized by a package of low-volatility assets denominated in four fiat currencies—US dollars, British pounds, euros and Japanese yen (and possibly the Singapore dollar). Libra aims at building a simple, borderless currency and serving billions of people with a financial infrastructure. The asset-side portfolio is designed much like the IMF's Special Drawing Rights (SDRs); meanwhile, very similar to traditional money market Fund (MMF). It is built on a blockchain network but is not fully decentralized. Rather, it is an alliance chain of multiple nodes (mainly financial payment institution).

economy, currency has experienced five stages, including physical money, weighed money, paper money, electronic money, and digital money. "The specific form of digital currency can be a number from a physical account, or a series of numbers credited to a name that is validated by specific cryptography and consensus algorithms. These digital currencies can be embodied or carried in a digital wallet, which can be used in mobile terminals, PC terminals or card bases. It is only electronic money with a normal number in a digital wallet, while it will be pure digital currency with a cryptographic number stored in a digital wallet and running on a specific digital currency network. E-money is simple in form and can be used with minor changes to the existing payment system, but it is largely dependent on the account system, less tamper-resistant, and more costly to know its customers and combat money laundering. Pure digital currency draws on the advanced technologies of various digital currencies today and thus is more difficult to tamper with and easier to use both online and offline with better visibility and wider range of channels. But it is more demanding in technologies and more difficult to operate and maintain, as it is based on a whole new ecosystem" (Jingdiwangtian et al. 2016). Digital currency, with different issuance and operation from previous currencies, performs all or some of the functions of traditional currencies, and functions more flexibly and intelligently in a certain field. Compared to paper money, digital money not only saves the cost of issuance and circulation, but improves the efficiency of transactions or investments and enhances the ease and transparency of economic transactions. However, "there is 51% of the threat of attacks on existing digital currencies despite the use of tight cryptography system", (Xiaoping 2016) with "black market" transactions, cyberattacks, account theft, and other incidents occurring from time to time. Security has become a major challenge in digital currencies.

The security of digital currencies has a critical impact on the economic and social development of each nation and even the whole world, which is evidently shown in each nation's attitude to Libra (Table 3.3). The G7,[11] which represents the world's largest economies, believe that Libra poses a risk to global financial system, and it should not be released for use before risks and challenges are fully eliminated. The G7 noted that even if Libra's supporters resolved the issue, the project may still not be approved by regulators. The reason is that Libra is an upgraded digital currency with functions and potentials of cross-border payments, super sovereignty currency, and new financial ecosystem. It will be constrained and challenged from four dimensions, i.e., technological innovation, commercial competition, regulation, and government (sovereign conflict). Specifically, if Libra is successfully launched and developed, it can disrupt global payment system in the short term, global monetary system in the medium term, and global financial market ecology and global financial stability system in the long term.[12] In the face of the global game of wisdom, technology, economy, politics, and power that Libra triggered, some nations have launched their

---

[11]G7 is a forum where the major industrial nations meet and discuss policies, including the United States, the United Kingdom, Germany, France, Japan, Italy and Canada.

[12]Zhu Min. The Possible Disruption of Libra. Sina.com. 2019, https://finance.sina.com.cn/zl/china/2019-09-23/zl-iicezueu7822307.shtml.

Table 3.3 Attitudes of Governments toward Libra

Nations	Attitudes	Concrete Manifestation
U.S	Prudent	The U.S. is not worried about the Libra shaking up its sovereignty legal currency status, but has been consistent in highlighting money laundering risks—President Trump talked about Libra in Twitter, arguing that bitcoin and other cryptocurrencies are "thin air", and Libra also has no standing or reliability. The U.S. has only one currency, and dollar was, is, and will be the strongest currency worldwide. Its Treasury Secretary Mnuchin revealed that before approving Libra, the U.S. would make sure very strict conditions were set to prevent Libra from being used for terrorist financing or money laundering
Australia		Australian regulators, currently pessimistic about Libra's prospects, said it faces a number of regulatory challenges. Governor of the Reserve Bank of Australia has said that Libra, launched by Facebook, will likely fail to be a mainstream cryptocurrency. He said that there are a lot of regulatory issues in Libra and Australian regulators will consider accepting it only if its stability is fully guaranteed
Japan		Governor of the Bank of Japan said Libra can have huge impact on financial system and needs global coordination for regulation, which should also include anti-money laundering policies. Meanwhile, Libra, as a cryptocurrency backed up by a package plan of legal currencies and government securities, will be difficult to regulate and pose a risk to the existing financial system. In this regard, Japanese authorities established a liaison conference group consisting of the Bank of Japan, the Ministry of Finance, and the Financial Services Agency to investigate Libra's impact on monetary policy and financial stability and to address its problems in regulation, taxation, monetary policy, and payment settlement
China		On July 9, 2019, Zhou Xiaochuan, former governor of the People's Bank of China and president of the China Finance Society, delivered a speech at the seminar themed "Reform and Development of Foreign Exchange Management in China", saying that Libra's attempt to peg a package of currencies represents the potential trend of a globalized currency. In response to this trend, he noted, "It is necessary and beneficial to study policies and identify potential risks in advance"
Germany	Opposing	Germany has made it clear in its Blockchain Strategy Draft published in June 2019 that it will not tolerate stable currencies like the Facebook-led Libra to pose threats on the nation's finance
France		Le Maire, French Minister of Economy and Finance, said, "I want to make sure Facebook's Libra won't become a sovereignty currency that can compete with national currency, because I will never accept corporations becoming private kingdoms"

(continued)

Table 3.3 (continued)

Nations	Attitudes	Concrete Manifestation
India		The Indian government has always been rejecting cryptocurrencies, and Libra's implementation is bound to be hampered. India's economic affairs secretary said, "Facebook has not yet fully explained the design of Libra, but in any case, it is a private cryptocurrency and we are not apt to accept it". In addition, the Indian government has recently introduced a draft bill banning cryptocurrencies, which will create a great hindrance to Libra's issue in India
Singapore	Neutral	Monetary Authority of Singapore (MAS) said there is insufficient information for them to decide whether to ban Facebook's cryptocurrency Libra or not
Thailand		Speaking at the Bangkok Fintech Expo on July 19, 2019, Santiprabhob, Governor of the Bank of Thailand said, "We won't be rushing into a decision on Libra. There come all kinds of new digital currencies, so the Bank of Thailand is monitoring all the new assets and is not showing any partiality for any particular financial service. The security of financial services is a top priority for banks, and that takes time"
Switzerland	Positive	Libra will help Switzerland play a role in an ambitious international project, said Swiss international finance secretariat
UK		Mark Carney, Governor of Bank of England said he would have an "open mind rather than a completely open door" to Libra

own digital currency research (Table 3.4). Laying out an anchor to whirlwind, China has made profound studies in digital currency for years and will launch its own digital currency DC/EP (Digital Currency/Electronic Payment) (Table 3.5). According to the Central Bank Digital Currency Survey published on January 23, 2020 by the Bank for International Settlements (BIS), among the 66 central banks surveyed, 20% are likely to issue digital currencies in the next six years, compared with 10% the same period last year. Referring to the historical evolution of currency, it is only a matter of time before the current monetary system is inevitably replaced.

There are some similarities in the attitude to digital currencies across nations, that is, all nations highlight cementing technological innovation and legal regulation. It can be seen from current digital currency that the supply side, demand side, and regulatory side still face many problems. The first problem is the "black market" on the supply side. Libra, though the most representative digital currency at present, is a private digital currency with no national credit endorsement and not considered a true currency (Table 3.6). Private digital currencies suffer problems, e.g., wide varieties, significant fragmentation, low market acceptance, and unpredictable technical risks, which endanger the sustainable development of digital currencies. Secondly, there is the "black market" on the demand side. Due to their convenience, concealment and other characteristics, digital currencies are apt to be used by criminals for fraud, illegal fund-raising, money laundering, and other illegal activities. For example, bitcoin, with its price unregulated, fluctuates greatly in value and is easily manipulated, which will bring huge financial losses to investors. Thirdly, there is

Table 3.4 Research of central banks on digital currencies in each nation (Region)

Nations (Regions)	Attitudes toward Issuing Central Bank Digital Currency (CBDC)	Reasons
Australia	No consideration for now	Failing to see the benefits of digital currencies over existing payment systems
New Zealand	No consideration for now	Unclear about the definitive benefits of a central bank issuing digital currency
Canada	Under research	Good CBDCs help online payment providers compete as cash becomes less competitive and alternative payments emerge
Norway	Under research	Can serve as a supplement to cash to "ensure people's confidence in currency and monetary system"
Brazil	Under research	To reduce costs in cash cycle; to increase efficiency and resilience of payment systems and money supply; capable of tracing and generating data; to promote digital society and financial inclusion
UK	Under research	Urgent need to keep pace with economic changes
Israel	Under research	To improve the efficiency of national payment systems; it can also be a monetary tool for central banks if interest is affordable; to help combat "shadow economy"
Denmark	Under research	To address the problems with paper money
Dutch Curaçao and Sint Maarten	Under research	It is costly and challenging for central banks to distribute funds between two regions, and digital currencies can make the currency union's payment system more secure and reduce the cost of anti-money laundering and real-name authentication
Singapore	Under research	The current digital currency SDGs within MAS's Ubin plan plays a role in inter-bank flows, but has not yet indicated a future opening to the public

(continued)

Table 3.4 (continued)

Nations (Regions)	Attitudes toward Issuing Central Bank Digital Currency (CBDC)	Reasons
China	Planning to issue	To reduce the cost of issuing and circulating banknotes; to improve the ease and transparency of transactions; to reduce regulatory costs; to increase the central bank's control over the supply and circulation of money
Sweden	Planning to issue	Can serve as a supplement to cash; to reduce the dependence of nationals on private payment systems and prevent failure of private payment system in times of crisis
Bahamas	Planning to issue	To improve the operational efficiency of communities, promote financial inclusion, reduce cash transactions, and lower the cost of services
Eastern Caribbean	Planning to issue	To reduce cash in circulation in the nation by 50%; to enhance the stability of financial sectors; to contribute to ECCU (European Chinese Joint Chamber of Commerce) members
Lithuanian	Planning to issue	Aiming to test cryptocurrency and distributed ledger technology
Thailand	Planning to issue	To speed up inter-bank transactions and reducing their costs with fewer intermediation processes
Japan	Planning to issue	To put Japan one step ahead in digital currency
Ecuador	Already issued	De-dollarization (unofficial statement)
Tunisia	Already issued	To promote the reform of domestic financial system
Senegal	Already issued	Financial inclusion
Marshall Islands	Already issued	To achieve national economic independence by replacing dollar currency circulation system

(continued)

Table 3.4 (continued)

Nations (Regions)	Attitudes toward Issuing Central Bank Digital Currency (CBDC)	Reasons
Uruguay	Already issued	High costs associated with opaque printing, distribution, transportation and transactions of banknotes
Venezuela	Already issued	The nation went into hyperinflation and original legal currency system collapsed

Source Blockchain Research Center, Institute of FinTech, Tsinghua University

also the "black market" on the regulatory side. The borderless online digital currencies lacking identifiable "issuers" pose challenges to effective regulation. On the one hand, "enforcement of regulation imposes costs on payment system providers and intermediaries, and these costs may ultimately be paid by the issuers or financial institutions responsible for issuing the digital currency. Some nations have begun to address the concerns of law enforcement by adapting existing regulations or developing new legal regimes" (Xiaowen 2016). On the other hand, the lack of regulation is an impediment to public confidence in digital currencies. Many participants may give up using digital currencies or investing in relative projects due to legal uncertainties or lack of protection. Globally, not many digital currencies are issued by sovereignty central banks, mainly because of the complexity and the wide coverage of risks, which will endanger economic and social security if there is no adequate study and precautionary measures.

3.11.2 Digital Currencies and Sovereignty Digital Currencies

Sovereignty digital currency, as a legal currency, is essentially the same as circulating paper money subject to national credit guarantees. It is featured by negotiability, storability, traceability, non-repudiation, non-falsifiability, controllable anonymity, non-repeatable transactions, and online or offline processing (Fig. 3.4). Sovereignty digital currency is more widely applicable than private digital currency and can help governments implement precise regulation. Negotiability refers to the fact that sovereignty digital currencies, as national legal currency, are endorsed with national credit and can be used as a means of circulation and payment for continuous flow of value in economic activity. Storability means sovereignty digital currencies take advantage of digitization and are securely stored in the form of data in the electronic devices of institutions and users for searching, trading, and management. Traceability means sovereignty digital currency transaction information consists of a data code and an identification code, the former specifying what is being transmitted and the latter specifying the source and destination of the data package. Non-repudiation

Table 3.5 Promoting digital currency by the People's Bank of China

Time	Events	Descriptions
2014	Study teams on issuing sovereignty digital currencies established	Formally demonstrating the feasibility of central banks issuing sovereignty digital currencies
2015	Digital currency study reports formed and two rounds of revisions completed on the prototype program for issuing sovereignty digital currencies	In-depth study of the issuance and business operation framework, key technologies, issuance and circulation environment, and legal issues of digital currencies, as well as its impact on economic and financial system, its relationship with privately issued digital currencies, and international experience in issuing digital currencies
January 2016	Digital currency seminar held	Stated for the first time that issuing digital currencies is a strategic goal of the central bank; highly appreciative of blockchain and other digital currency technologies, indicating that it will actively study and explore the possibility of central banks issuing digital currencies
July 2016	The joint research group of "Operation and Supervision of Digital Currency and Currency-like Digital Assets" with the participation of National Development and Reform Commission (NDRC) launched in Beijing	Indicating a two-year systematic study on establishing a governmental or public regulatory mechanism related to sovereignty digital currencies
January 2017	Digital Currency Institute officially established in Shenzhen	In order to test and experiment with blockchain technology in real life, to study blockchain and digital currencies, so as to ensure blockchain technology can be applied to the fullest extent to China's financial sector
February 2017	Successful testing of blockchain-based digital bill trading platform	Legal digital currencies issued by the central bank having been piloted on the platform
March 2017	Central bank deploys fintech efforts and builds innovation platform led by digital currency exploration	The goal of building an innovative platform which started with digital currency emphasized on the technology conference of People's Bank of China
May 2017	Central Bank Digital Currency Research Institute officially opened	Yao Qian as director, whose research interests include digital currencies and FinTech

(continued)

Table 3.5 (continued)

Time	Events	Descriptions
January 2018	The experimental production system of digital bill trading platform successfully put into operation	Based on the frontier of blockchain technologies and actual bill business, the system carried out a comprehensive transformation and improvement of the platform prototype system of preliminary digital bill trading and assisted Industrial and Commercial Bank of China, Bank of China, Pudong Development Bank and Bank of Hangzhou in successfully completing the issuance, acceptance, discounting, and rediscounting of digital bills based on blockchain technology
August 2018	Nanjing FinTech Research and Innovation Center inaugurated	Co-founded in Nanjing University by Nanjing Municipal People's Government, Nanjing University, Bank of Jiangsu, People's Bank of China Nanjing Branch and People's Bank of China Institute of Digital Currency. Focusing on FinTech and financial service innovation, it provides financial innovation for the construction of Innovation City by building a linkage among government, industry, academia, research, and application
September 2018	The Central Bank Digital Currency Institute built the PBCTFP	The Central Bank Digital Currency Institute and the Shenzhen Branch of the People's Bank of China promoted the establishment of PBCTFP to help alleviate the financing problems of small and micro enterprises in China, and to build an open financial and trade ecology based in the Guangdong-Hong Kong-Macau Greater Bay Area for China and the whole world
May 2019	PBCTFP developed by Central Bank Digital Currency Institute unveiled	PBCTFP has been implemented to serve trade and finance in the Guangdong-Hong Kong-Macao Greater Bay Area

(continued)

Table 3.5 (continued)

Time	Events	Descriptions
July 2019	The central bank will promote digital currency R & D	Wang Xin, director of the central bank's research bureau, said at the launch of the Digital Finance Open Research Program and the First Academic Research Conference that the central bank will promote digital currency R & D in the future
August 2019	Central government supports digital currency research in Shenzhen	*The Opinions of the Central Committee of the Communist Party of China and the State Council on Supporting Shenzhen in Building an Early-Start Demonstration Zone of Socialism with Chinese Characteristics* clearly states that it supports innovative applications, e.g., digital currency research and mobile payments in Shenzhen
Sept. 2019	Part of central bank's paper money will be replaced by digital currency	The People's Bank of China, which integrates digital currencies and electronic payment instruments, plans to launch a package aimed at replacing some of the cash, central bank governor Yi Gang said at the first press conference at the press center for celebrating 70th anniversary of the People's Republic of China
	Central bank plans to launch digital currency	Huang Qifan, vice chairman of China Center for International Economic Exchanges (CCIEE), said at the first Bund Finance Summit that the central bank will launch digital currency

N.B. The above are incomplete statistics

means sovereignty digital currency uses security technologies such as digital time stamps to make sure both parties of the transaction cannot repudiate the transaction and the various elements of the conduct that have taken place. Non-falsifiability refers to the fact that various security techniques such as hashing algorithms are used in creating and issuing sovereignty digital currencies to prevent it from being illegally copied, counterfeited, and altered. Controllable anonymity means real name registration is voluntarily needed on the front and necessarily needed on the back, so that no participant other than the monetary authority is able to know the identity of the owner or previous user. Non-repeatable transaction means the owner of sovereignty digital currency may not pay it to more than one user or merchant in succession or simultaneously, which tackles double spending. Online and offline processing means it can be transacted electronically without a direct connection to a host or

Table 3.6 Comparison of DC/EP with Libra

Items	People's Bank of China	Facebook
Digital currency	DC/EP	Libra
Research time	2014	2018
Coinage rights	I/A	N/A
Focus	Payment	Payment
Problems to be addressed	Modern dilemma of currency	Current financial services deficit
Blockchain types	Alliance chain	Alliance chain
Anchor	Value of *Renminbi* or digital assets	US dollars or a basket of other currencies
Anonymity	Non-anonymity traceability	Non-anonymity
Clients	Users in China	Over 2 billion users on Facebook

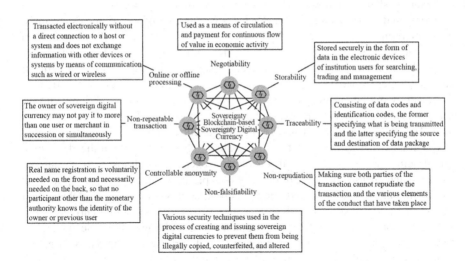

Fig. 3.4 Sovereignty digital currency based on sovereignty blockchain

system and does not exchange information with other devices or systems by means of communication such as wired or wireless.

For sovereignty states, the establishment of sovereignty digital currencies will have significant implications for their position in global economic system and is also a major way for them to establish dominance in the financial innovation. There has been a dramatic change in payments globally, and the rise of digital currencies has brought new opportunities and challenges to central banks' currency issuance and monetary policies in all nations. Seeking innovative support for payment systems via the issuance of digital currencies and mastering the opportunities of blockchain technology is now an important measure for global international financial centers to consolidate their positions. Judging from the status quo and use of digital currencies, blockchain technology-based digital currencies only tackle the credit problem and are potential to be a financial product and asset, but they will never become a good currency if they cannot adjust currency fluctuations and if there is no supply adjustment mechanism adapted to economy. In addition to being a medium of exchange, sovereignty digital currencies can also function as a standard of value, store value, and provide deferred payments service. It also plays a role in purchasing power hedging, advancing credit economy, transforming payments and remittances, equity clearing and settlement, equity crowdfunding, note business, audit reform, etc. That is why all nations are actively exploring sovereignty data currencies. Meanwhile, digital currencies based on blockchain technology have made the transaction process traceable, but whether it will become a currency still depends on the participants' recognition and its value stability, while the key is that its payment and clearing must meet regulatory requirements. Sovereignty digital currencies can better integrate with existing monetary system, which will accelerate digital currencies globally.

Compared with sovereignty currencies, sovereignty digital currencies are innovative in terms of technology and governance, and will have a significant impact on credit mechanisms, policy regulation, financial innovation, etc. "Monetary theory holds that sovereignty currency is a credit issued by the state, based on national taxation, safeguarded by law, liquidated and guaranteed by national taxation".[13] Sovereignty digital currency not only has the characteristics of sovereignty currencies, it also has distinct advantages. First, it has lower cost and more efficient circulation. The trust mechanism of sovereignty digital currencies is based on asymmetric cryptography, and users conduct trustable value exchange in a decentralized polycentric structure, where the frictional cost of value exchange is almost zero, because its payment and clearing are done directly by both parties without third-party payments or clearing agencies. Particularly, in the current context of economic globalization and global trade acceleration, the application of sovereignty digital currencies has strong economic value due to the scale and frequency of transactions. Second, it can effectively deal with currency over-issuance and curb hyperinflation. As a nation controls the issuance of currency, it is inevitable that banknotes will be over-issued for the purpose of reducing fiscal deficits and developing economy, which is subject to hyperinflation, causing damage and erosion to social wealth and disrupting smooth economic development. Sovereignty digital currency is issued on defined transactions or specific needs recognized by various participants and is a direct embodiment of the transaction and the value of goods and services throughout society. It is a true reflection of economic development, so theoretically, there will be no over-issuance and no inflation (Project team of Yibin Central Sub-branch of the People's Bank of China 2016). Third, it can improve the accuracy and effectiveness of monetary policy. Sovereignty digital currencies are nominal currencies, and their tamper-resistant and

---

[13]Han Yuhai. Monetary Sovereignty and the Fate of the State. *Green Leaf*, 2010, Z1, p. 155.

non-falsifiable time stamps reflect complete details of transactions and information on both parties. It can record the transaction credit of each participant faithfully and form a uniform ledger throughout the system. Thus, national regulators can have a comprehensive and accurate first-hand understanding of the implementation of monetary policy, credit policy, and national industrial policy through tracing the ledger information and the circulation of sovereignty digital currencies, so as to assess the effects of policies, adjust and optimize relevant policies according to the situation. Fourth, it will help build a sound and efficient new financial system. In the era of big data, finance and big data are integrating deeper, and digital innovation of the financial system is on the rise. The users and institutions of private digital currencies, such as Bitcoin and Ripple, are steadily expanding, which brings multiple impact on the penetration and diversion of financial systems in sovereignty states. "The central bank's digital currency adopts a new payment system and mode that supports 'peer-to-peer' payment and settlement, with a reduction in currency transaction intermediation. Currency circulation network would be greatly delayered, the interconversion of financial assets accelerated, and the efficiency of transactions significantly improved. More importantly, sovereignty digital currencies adopt transparent bookkeeping and controllable anonymous transactions, which eventually leads to a large, sophisticated and transparent network of big data, where sovereignty central banks use data to monitor and assess the risks in the financial system and ultimately build a robust and efficient new financial system" (Xun 2017).

3.11.3 Sovereignty Digital Currency Under Sovereignty Blockchain

Sovereignty blockchain, driven by both technology and system, makes sovereignty digital currencies possible. It is a secure distributed ledger technology with rules and consensus at its core under national sovereignty and national law and regulation. It is not only the use of a range of new technologies, but more importantly an innovation at the institutional and regulatory level, which is regulable, governable, trustworthy, and traceable. Sovereignty digital currency is an important scenario for the application of sovereignty blockchain. The direction of current financial innovation is sharing, which needs more transitional reforms with the aim of addressing existing conflicts in economic and financial operations. Sovereignty blockchain is based on blockchain, proof of workload and proof of entitlement to delve into right technologies and rules, so it better meets digital monetary system. It aims at going beyond technological revolution to systemic revolution and fundamentally alter the traditional organization, management, information transmission, and resource allocation to realize a "stable, orderly, optimal and balanced" sovereignty monetary system. A sovereignty digital currency built and issued on the sovereignty blockchain has the advantages of "digitalization" and "centralization". Such a system is unlikely to adopt a fully decentralized cryptographic digital currency model. Rather, it requires a completely innovative hybrid technological architecture to support it. With this technology, the transmission and circulation paths of encrypted numbers can be fully documented, stored, and shared by distributed ledgers, making the path of sovereignty digital currency flows traceable and tamper-resistant with traceability and non-repudiation. Unlike private digital currencies which are completely decentralized, sovereignty digital currencies use a decentralized polycentric network structure with the support of national credit and is guaranteed, signed and issued by sovereign central banks, so it has the unique advantage of centralization, ensuring more stable pricing and greater appeal for social recognition.

Sovereignty digital currency, essentially a sovereignty currency, helps build an architecture of "one currency, two banks and three centers" (Fig. 3.5). The "two banks" refers to the fact that sovereignty currency follows the binary model of "central bank-commercial bank" (Fig. 3.6), which means sovereignty central bank issues the currency to commercial banking pools, then commercial banks, entrusted

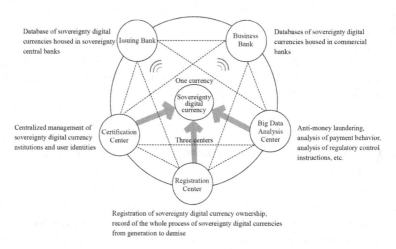

Fig. 3.5 The architecture of "one currency, two banks, three centers" of sovereignty digital currency

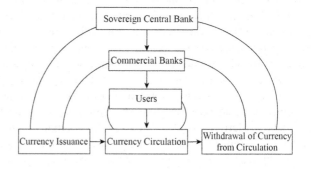

Fig. 3.6 The binary model of "central bank-commercial bank" of sovereignty digital currencies

by sovereignty central banks, provide the public with access to sovereignty digital currency and other services, and work with the former to maintain the operation of sovereignty digital currency issuance and circulation system, during which only the transporting and storage are changed. On this basis, three centers of certification, registration, and big data analysis are included, thus an architecture with the core elements of "one currency, two banks, three centers" is formed. The certification center is a fundamental component of system security, enabling sovereignty central banks to carry out centralized management on sovereignty digital currency institutions and user identities. It is also a key part of controllable anonymity design. The registration center records sovereignty digital currency and corresponding user identity, enables ownership registration and flow record, and completes the registration of the entire process of generation, circulation, inventory verification, and demise. The big data analysis center is a monitoring module to secure transactions, prevent illicit transactions, and enhance the effectiveness of monetary policy. According to business needs, it can analyze kinds of trading behaviors in real time, which helps regulators realize data analysis and accurate decision-making.

3.12 Digital Identity

Identity is a status enjoyed by all members of the community, and all those who hold that status are equal in terms of rights and obligations. Society is an unequal system. Social class, like citizenship, can also be built on a set of ideals, beliefs, and values. Thus, it is reasonable to assume that the impact of citizenship on social class will take the form of conflicts between two opposing principles (Zhonghua and Xunlian 2017). Existing identity system hinders technological innovation and constrains the digitization, networking, intelligence, pluralism, and collaboration of social services, while the digital identity creates conditions for bridging the digital divide and sharing the material and cultural-ethical fruits of economic development among all citizens.

3.12.1 From Traditional Identity to Digital Identity

"Humans, as social creatures, have developed a special concept in their interactions with each other—identity. Identity is a symbol of the differences in status between people, and to some extent also reflects the differences in behavioral capacity between different people".[14] Specifically, "At the social level, identity refers to a person's position in a certain social system, or a person's social position when interacting with others in social life" (Guoqiang 1990). At the legal level, identity refers to the sum of the individual capacities within legal framework, i.e., the sum of individual legal

---

[14]Liu Ruxiang. *Person, Identity, Contract: on Theory of Law of Society* written by Dong Baohua, et.al. Edited by Su Li: *Law Book Review*. Law Press · China, 2003, p. 79.

rights and obligations. It is rational to say that all the forms of "identity" mentioned in "human law" originate from the ancient rights and privileges belonging to a "family". "The individual does not create any rights or obligations for himself. The rules which he is subject to derive, first, from the place of his birth and, secondly, from the imposed orders given by the head of the household of which he is a member".[15] Further, the distribution of rights and obligations is determined by people's status (noble or commoner, father or son, husband or wife, etc.) within a "particular group" such as a family.

In the "status society", identity (birth) is the main approach by which people acquire privileges. "Human flesh enables him to become the bearer of a particular social function. His flesh becomes his social right".[16] Identity becomes the fundamental criterion for determining the status, rights, and obligations of people. The essence of identity is about differences, affinity, superiority, and inferiority. Thus, identity becomes the source of all the differences between people. In the meantime, identity is also the fundamental criterion for the allocation of power, because power is derived from and differentiated by identity. There is no power without identity. Different identities make power more unequal and privileged. In a "status society", identity is equated with power and authority, which stimulates people to worship power, idols, and status, and to indiscriminately respect and obey orders from people with higher status. Therefore, the "status society" is a society ruled by men. In such a society, all people belong to different classes according to their origin, property, and occupation, and each class has its own views, feelings, rights, habits, and ways of life. People from different classes are too different in their thoughts and feelings to make others believe that they are of the same nationality (Tocqueville 1990).

With the progress of digital technology, digital identity has emerged and may become an important part of future blockchain infrastructure. "Digital identity is a new type of identity formed in the context of modern computer technology based on modern communication and network technology" (Jingyu and Zhihong 2018). The Internet era and the future blockchain era share the feature of "digitalization", the basis of which is the digital identity of the user. However, "with the growing Internet and the soaring number of websites, centralized identities have brought about a lot of confusion and limitations. Users' digital identities are scattered randomly across the Internet together with associated data, which are not under users' own control, leading to the risk that individuals' right to privacy simply cannot be effectively safeguarded".[17] A digital identity system built via blockchain technology ensures the authenticity and security of data and effectively protects user privacy. However, for the absolute security of digital identities, a complete solution integrating software and hardware need to be formed with a variety of digital technologies complementing

---

[15] [UK] Maine. *Ancient Law*, translated by Gao Min, Qu Huikhong. China Social Sciences Press, 2009, p. 176.

[16] [Germany] Marx, [Germany] Engels. *Marx and Engels Collected Works (Vol.1)*, translated by Central Compilation and Translation Bureau. People's Publishing House, 2006, p. 377.

[17] TokenGazer and HashKey. Decentralized Identity (DID) Research Report. TokenGazer, 2019, https://tokengazer.com/#/reportDetail?id=240.

each other and working together. It was suggested on the 2018 World Economic Forum in Davos that a good digital identity should meet five elements. "First, reliability. A good digital identity should be reliable to build trust in the people it represents and enable them to exercise rights and freedoms, thus proving they are eligible for service. Second, inclusiveness. Digital identities can be created and used by anyone who needs them, free of discrimination based on identity-related data, and without the need for identity verification that excludes identity. Third, usefulness. It is easy to create and use useful digital identities, which provide access to multiple services and interactions. Fourth, flexibility. Individual users can choose how their data is used, what data is shared for which transactions, with whom, and for how long. Fifth, security. Security means it can protect individuals, organizations, or devices from identity theft and misuse to avoid unauthorized data sharing and human rights infringement" (Qianren et al. 2019).

3.12.2 Digital Identity and Destratification

With the development of digitalization, networking, and intelligence, digital technology has become a new variable reshaping the mechanisms of social stratification. In social stratification studies, there is a "modern-postmodern" approach, which is actually a "hierarchical–de-hierarchical" model of analysis. From the perspective of stratification, the digital divide can be regarded as the result of "Matthew effect" of the information age, which means "digital divide is a kind of 'technological divide', i.e., the advanced results are not shared fairly, thus creating a situation where "the rich get richer and the poor get poorer" (Renzong et al. 2014). In other words, through the effective transformation of data capital into other capital, the social and economic gap between social classes is widening and eventually reinforces different class positions. This is a structuralist idea that social class structures are continuous and that digital technology is a new variable facilitating stratification. In contrast, postmodernist perspective stresses the importance of culture and the concept of individuals. It emphasizes the construction of individuals on their own and holds that pluralistic lifestyles and consumption practices will have a tremendous impact on the established hierarchical structure. Thus, it can be seen that digital technology can help individuals break loose the constraints of inherent hierarchical structure and realize social hierarchy "destratification" (Sheng 2006).

Traditionally, social classes are divided by a single dimension or simple multi-dimensional integration of economy, politics, society, culture, and honor. But in the digital world, the division of social classes relies on digital dimension with information and communication technology as the core, which mainly includes digital awareness, ICT (Information and Communication Technology) access and use, information contents use and creation, digital information qualities and digital cohesion.[18] Based

---

[18]Digital cohesion is the degree to which members of a social group are united in cyberspace using information and communication technologies and digital information. It is manifested in the capacity

Table 3.7 Qualitative description of classes in digital society

Classes in digital society	Characteristics
The digital elite	Have motivation and will; have access to and use information resources; digitally literate; can create and upload digital contents; be able to access and use ICT; digitally cohesive
The digital affluent	Have motivation and will; have access to and use information resources; digitally literate; can create and upload digital content; be able to access and use ICT; lack of digital cohesion
The digital middle class	Have motivation and will; have access to and use information resources; digitally literate; do not create and upload digital content; be able to access and use ICT; lack of digital cohesion
The digital poor	Material poverty: have motivation and will; do not have access to information resources; be digitally literate; do not create and upload digital content; not able to access ICT; lack of digital cohesion
	Consciousness poverty: lack motivation and will; do not have access to information resources; not digitally literate; do not create and upload digital content; be able to access ICT; lack of digital cohesion
	Literacy Poverty: have motivation and will; do not have access to information resources; digitally illiterate; do not create and upload digital content; be able to access ICT; lack of digital cohesion
The extreme digital poor	lack motivation and will; do not have access to information resources; digitally illiterate; do not create and upload digital contents; not able to access ICT; lack of digital cohesion

Source Hui (2013)

on this and the theoretical and analytical approach of social stratification, communitarianism theories and the performance dimension of digital inequality, communities and members in the digital age are divided into five levels, namely, digital elite, digital affluent, digital middle class, digital poor, and extreme digital poor (Table 3.7). The digital elite is the only group among the five that has digital cohesion. The digital affluent group is characterized by being able to achieve digital affluence by creating and uploading digital contents. The digital middle class is featured by having access to basic information and communication devices and being equipped with digital awareness, digital literacy, and motivation and desire to use computers and Internet. They get online information passively with ICT, but do not necessarily use these resources to solve practical problems. The digital poor is defined as people who fall into one or more of the categories of material poverty, awareness poverty, and literacy poverty in terms of ICT and information content. The extreme digital poor is the sum of the three phenomena of digital poverty (Hui 2013).

---

to appeal to online interests, to maneuver online public opinions, and to influence decision-making in practice. The two extremes of cohesion can be described by "unity" and "fragmentation". "Unity" represents digital cohesion, while "fragmentation" is defined as a lack of digital cohesion. Accordingly, only the digital elite is united, while the other four strata are considered the "fragmented" digital class.

Digital identity offers the possibility of bridging digital divide. Amartya Sen, Nobel laureate in economics, has noted that, "In a nation where information flows freely, there is no real poverty. The openness, circulation, and enjoyment of information is conducive to the balanced and sound development of economy and society". In light of the causes, specific manifestations and ethical crisis of the digital divide, the spirit of sharing, contracting, and humanism should be promoted in the digital era. First of all, in the digital age, mastering data means having a resource advantage and absolute dominance in productive life. Therefore, to eradicate this inequity, it is necessary to eliminate data fragmentation and data isolation, which requires the full exploitation of the reliability, inclusiveness, usefulness, flexibility, and security of digital identity to lay a solid foundation for protecting data privacy and promoting data sharing, openness, and circulation. Second, data sharing will inevitably lead to disputes over rights and obligations, interests and responsibilities among stakeholders. If these disputes are not dealt with properly, the spirit of sharing will eventually be reduced to lip service. Therefore, there is a need to clarify the boundaries of rights and obligations, interests and responsibilities among big data stakeholders from the perspective of contractual ethics (Shiwei 2018). Finally, "big data technology is not omnipotent and cannot solve all problems. It is only a quantitative means of decision-making. A proper understanding of the rights and wrongs of objects and following human spirit is a much more important prerequisite" (Jianming 2013). Digital technology has brought great economic value as well as humanistic value. Only by actively promoting the humanistic spirit and tapping into humanistic values can the digital age be potential to be a more equal and harmonious era to overturn the traditional ways of living.

3.12.3 Blockchain-Based Digital Identity

Multipolarization, economic globalization, and social informatization today are soaring, and global economic governance system is being restructured at an accelerated pace. Opportunities and challenges exist in the midst of these changes, and it is crucial to improve sci-tech innovation capacity and accelerate economic structure and industrial chain adjustment. The emergence of digital identity is changing the drive and mode of economic society. Compared to traditional identity systems, it will significantly improve overall social efficiency and unlock maximum user value, benefiting all parties, including government, service providers and users (Yihui 2019). Digital identities are important, and vulnerable as well. In practice, the security of digital identities faces considerable threats. Judging from the existing regulatory measures on digital identity protection in different nations, personal data protection regulations and privacy regulations are the legal basis for digital identity, and digital identity has not been as widely protected as the right to privacy (Gang et al. 2015). While all nations are tightening personal data controls, a series of pain points in digital identity, such as information fragmentation, false data, easy leakage of data, and the difficulty of user self-control, need to be effectively addressed.

Blockchain, by virtue of its technical features, provides a relatively credible solution to the trust verification and independent authorization of digital identity. More typical are identity chain management systems and digital identity systems based on blockchain technology. "According to Research & Markets, global blockchain identity management market is forecast to grow from $90.4 million in 2018 to $1.9299 billion in 2023" (China Academy of Information and Communications Technology 2019). Identity chain leverages the features of blockchain and achieves "trustworthy data and more trustworthy people". "Identity chain is a unique marker for network electronic identity authentication based on blockchain, which can protect identity information from being leaked, expand the service and scope of network identity coding by utilizing the features of blockchain, thereby improving the service capability of network identity coding, providing various types of application systems with secure, reliable and diverse forms of trusted identity authentication services, and protecting user privacy. The digital identity chain correlates and carry out unified management on the offline physical identities to online digital identities by giving all participants digital identities to make data open and transparent. Users decide whether to open or partially open their data and who has access to open data through permission settings, so data is not manipulated by others and users' privacy is well protected" (Junsheng et al. 2019).

A digital identity system based on blockchain ensures the digital identities, and the data generated in a series of associated activities and transactions are authentic and valid. Studies have shown that all identity systems share something in common— they are made up of elements of users, identity providers, and relying parties.[19] Digital identity systems are no exception, with its sophisticated identity screening techniques and security protocols ensuring that identity records are difficult to be destroyed, tampered with, stolen, or lost, and they can be divided into five basic types, namely, internal identity management, external verification, centralized identity, unified verification, and distributed identity (Table 3.8). And a successful natural identity network should follow five main principles. First, social value. The identity system should be able to be used by all users and maximize stakeholder benefits. Second, privacy protection. User information shall be provided only to the right entities in right circumstances. Third, user-centeredness. Users should be able to control their information and decide who has the right to hold and access it. Fourth, viability and sustainability. Identity systems should be a sustainable business with high resilience to political volatility. Fifth, openness and flexibility. Identity systems should be built on the standard of openness to ensure they are extensible and developable, and systemic standards and guidelines need to be transparent to stakeholders (Deloitte 2017).

---

[19] A user is a person who has an identity within the system that allows a transaction to take place. An identity provider is a person who stores a user's attributes, ensures the authenticity of information and completes transactions on behalf of the user. A relying party is a person who serves as a user after the identity provider has provided the user with a guarantee.

Table 3.8 Overview of digital identity systems

Types	Relationship between identity providers and relying parties	Characteristics
Internal identity management	The entity is both an identity provider and a relying party	Best suited to managing user rights based on internal information within a single entity to ensure the right people have access to right resources
External verification	Multiple identity providers authenticating users for a single relying party	Best suited to simplifying a range of services provided by a single entity without requiring users to log in repeatedly for different services
Centralized identity	One identity provider serving multiple relying parties	Best suited to providing different users complete, accurate, and standardized non-private data
Unified verification	A set of identity providers verifying users for multiple relying parties	Best suited to providing complete, accurate, and standardized user data for multiple entities, where the users have access to services provided by multiple entities without requiring duplicate logins
Distributed identity	Multiple identity providers serving multiple relying parties	Best suited to providing users with a convenient, controlled, and privacy-protected service in a networked environment

Source Deloitte (2017)

3.13 Digital Order

There are various structures in nature and human society, and orders, formed with the arrangement and combination of structures, have specific functions and roles. While promoting human social patterns, they have influenced the way of human life (Tingxiao and Xuan 2013). David Weinberger in his book *Everything Is Miscellaneous* divided order into three types. The first is physical order, referring to the structure of existence of physical things themselves. In this sequence, objects are limited by their own space and time, and follow the logic of "one object belongs to one position" according to some kind of arrangement logic in a fixed physical space. The second is rational order, which is artificial and virtual. In this sequence, objects are separated from, yet connected through certain reference to their own information. Information becomes the agent object of the first order and points to the physical location of the object by encoding. The third order, or digital order, is a specific new order that has been created in the digital age to meet individual needs. In the digital environment, this order is changing the way we produce and live (Weinberger 2017).

Connection, trust, and identity are the three elements of digital existence. Individual digital citizens connect with each other via technological tools and gradually develop trust and identity for achieving common goals. Nicholas A. Christakis thought, "At the heart of civilized society lies the need for people to build connections with each other. These connections will curb violence and become the source of comfort, peace and order. Rather than loners, people become collaborators" (Christakis 2012). The digital society is seen as the "second living space" for human life and practice, which is a more figurative expression than "network society" or "virtual society". The rapid development and wide application of digital technology has given birth to digital society, which is a distinct technological and social architecture and special socio-cultural form. "Advancing digital technology and digital society have become the key features of contemporary human social evolution, and this process is inherently inevitable and irreversible" (Yi 2019). In the future digital society, people are connected in a vast social network through digital identities and digital currencies. Such interconnectivity is not only an innate and essential part but an eternal force of our lives.

The digital society, driven by digital identity, digital currency, and many other factors, presents a structure and operation state that is different from the past. In terms of operational state, the digital society is characterized by the following four features. The first is cross-domain connectivity and full-time presence. Cross-domain connectivity primarily achieves universal connectivity.[20] Relying on the unique convenience of virtualization brought about by digitalization, it provides a revolutionary solution to transcending geospatial constraints and achieving effective connectivity so as to truly achieve the interconnectivity and integration of global networks. The second is autonomy of action and in-depth interaction. The digital society, the network era, and cyberspace provide extremely convenient conditions for the free activity of digital citizens. It not only enables the virtual representation of human cyber-behavioral activities, but allows these activities to continue in cyberspace, so digital citizens can interact more deeply with each other. The third is data sharing and resource integration. The resource integration in cyberspace transcends the geographical and spatial boundaries of reality, and thus easily and quickly docks and combines resource elements and improves the effectiveness and timeliness of resource integration and utilization. The fourth is intelligent control and efficient collaboration. Mechanization, automation, and intelligence are the welfare of human society brought by sci-tech progress. A series of intelligent and automatic equipment are providing people with convenient and efficient services. The network realizes both the connection of objects and people, so Internet is more than a technology or a tool. What hides behind are the connectivity and relationship network in social and cultural senses. Consistent with the integration of resources in cyberspace, people, relying on cyberspace as a

---

[20]Universal connectivity includes both digital connections between humans, digital connections between objects such as smart devices, as well as digitalization-based connections and penetrations between humans, objects, and smart devices.

platform and field, can cooperate with each other in different areas of work and life (Yi 2019).

The opportunities offered by digital technology are evidently visible, but the corresponding risks are also looming larger. Despite rapid technological advances, efficient forms of innovation and a higher quality of life they have brought, digital technologies have also caused "creative destructions", problems and challenges to people's lifestyles, social structures, modes of governance, states of order, etc., which puts the institutional system, operating mechanism, regulatory approach, and social order in the possibilities of being "disrupted" and "reconstructed" (Changshan 2018). On December 10, 2018, the report of *Our Shared Digital Future: Building an Inclusive, Trustworthy and Sustainable Digital Society* was released on the World Economic Forum, providing a new vision for addressing the issues and challenges posed by digital technology. The report argues that existing mechanisms and structures are

Table 3.9 Six common development goals for a shared digital future

Common development goals	Objectives	Barriers	Cooperation focuses
Leaving no person behind	Every person can access to the Internet	• There is a geographical divide in economic cooperation and development organization • There are gaps between groups of different incomes, ages and genders in some nations	• Smart fiscal policy • Effective spectrum allocation • Adopting or revising national broadband plans • Making sustainable digitalization an overreaching goal for the government • Unlocking the potential of universal service and access funds • Incorporating diversity objectives in strategies and investments • Platforms for scaling leading practices and bottom-up innovation
Empowering users through good digital identities	Every person can access digital services through identities that are secure, effective, useful, usable, and offer real choice	• Geographical divide in digital identity • Gender divide in digital identity • Lack of standards for "good" digital identity	• Defining what "good" looks like: • Promoting common digital identity standards • Building awareness and capacity • Supporting and amplifying technology innovations • Developing policy toolkits • Platforms for learning and action

(continued)

Table 3.9 (continued)

Common development goals	Objectives	Barriers	Cooperation focuses
Making business work for people	Digital enterprises create sustainable value for all stakeholders	• Disruption • Technology, responsibility, and trust • Platforms and networks • New technology architecture • Planetary boundaries of the digital environment	• Encouraging a network for responsible business leadership • Developing tools and guides on transformation • Empowering boards and companies' teams • Private–public cooperation to develop shared strategies for industry transformation • Private–public collaboration on important digital topics
Keeping everyone safe and secure	Everyone's identity, assets, reputation, and life are protected from cyber-risks through trusted and secure technologies, enterprises and institutions along with a culture of cybersecurity and cyber-resilience	• To stop global cyberattacks and curb cybercrime • Leadership responsible for setting strategy assumes strategic responsibility for network resilience • Not a single nation or business can deal with cybersecurity issues alone	• Building human capital • Building understanding at leadership level • Addressing confusion and fragmentation • Encouraging technical innovation • Mitigation strategies • Intelligence sharing • Global capacity-building and training programs • National and corporate strategies
Building new rules for a new game	We have an effective set of rules and rule-making tools for the Fourth Industrial Revolution that are agile, inclusive and multistakeholder-ready as needed	• Different strategic priorities of various stakeholders • Industry-led mode leads to problems regarding industry and market mechanisms, super-regulators, ethical standards, transparency of technological innovation, etc • Technology-based innovation	• Accelerating global capacity for inclusive agile-governance development • Establishing policy laboratories within government • Using regulatory sandboxes to encourage innovation • Increasing agility through the use of technology • Promoting governance innovation • Crowdsourcing policy-making • Promoting collaboration between regulators and innovators • Sustaining progress

(continued)

Table 3.9 (continued)

Common development goals	Objectives	Barriers	Cooperation focuses
Breaking through data barriers	Common practices, tools, and resources that allow us to benefit from data while protecting the interests of all stakeholders	• Enterprises are not willing to expose proprietary or commercially sensitive information • Facing with a variety of legal environments • Many industries are subject to stricter regulation	• Exploiting interest in development, health, environment and humanitarian sectors • Rapid prototype of these leading models • Exploring development of a common and consistent risk-based framework • Creating and sharing legal agreement templates • Supporting policy-makers with insights and tools • Develop suite of technical, legal, and market approaches

Source Xu Jing. Our Shared Digital Future: Building an Inclusive, Trustworthy and Sustainable Digital Society. *The Internet Economy*, 2019, No. 5

somewhat overstretched to cope with the challenges of digital future, which must be inclusive and socially, economically, and environmentally sustainable. In order to shape such a future, governments, business, academia, and civil society need to share common goals and concerted actions. Global leaders and organizations should focus on the six shared development goals, namely, leaving no person behind, empowering users through good digital identities, making business work for people, keeping everyone safe and secure, building new rules for a new game, breaking through data barriers, for better communication and cooperation, and a common platform to shape our digital future (Table 3.9).

References

Yang Baohua, Chen Chang. *Blockchain: Principles, Design and Applications*. China Machine Press, 2017.

Han Bo. Acquaintance Society: A Possible Way to Construct Internet Integrity in the Big Data Era. *Social Sciences in Xinjiang*, 2019, No. 1.

Bofeng, Chen. 2011. On the acquaintance community: an ideal-type exploration of the mechanism in the village order. *Chinese Journal of Sociology* 1.

Ma Changshan. Message to the New Column. *ECUPL Journal*, 2018, No. 1.

China Academy of Information and Communications Technology. Blockchain White Paper (*2019*). China Academy of Information and Communications Technology website. 2019.

Christakis, Nicholas A., and James H. 2012. *Connected: The amazing power of social networks and how they shape our lives*. Translated by Jian Xue. US, China: Renmin University Press.

Deloitte, World Economic Forum. Picture Perfect: A Blueprint for Digital Identity. Deloitte. https://www2.deloitte.com/cn/zh/p.s/financial-services/articles/disruptive-innovation-digital-identity.html.

[US] Thomas Friedman. *The World is Flat: A Brief History of the 21st Century*, translated by He Fan, Xiao Yingying, Hao Zhengfei. Hunan Science and Technology Press, 2008.

[US] Francis Fukuyama. *The Great Disruption: Human Nature and the Reconstitution of Social Order*, translated by Tang Lei. Guangxi Normal University Press, 2015.

Xie Gang et al. On Digital Identity and Protecting Measures of Electronic Public Service in Big Data Era. *Forum on Science and Technology in China*, 2015, No. 10.

Zheng Ge. Blockchain and Law in the Future. *Oriental Law*, 2018, No. 3.

[UK] Anthony Giddens. *Modernity and Self-Identity*, translated by Zhao Xudong and Fang Wen. SDX Joint Publishing Company, 1998.

[UK] Anthony Giddens. *The Consequences of Modernity*, translated by Tian He. Yilin Press, 2011.

Yang Guangfei, Friendvertising: A Face of the Transmutation of Interpersonal Relationships in Transitional China. *Academic Exchanges*, 2004, No. 5.

Wu Guanjun. The Cunning of "Trust"—Rethinking Trust in the Trust-lacking Age. *Exploration and Free Views*, 2019, No. 12.

Hao Guoqiang. From Personal Trust to Arithmetic Trust: Study of Block Chain Technology and Construction of Social Credit System. *Journal of Nanning Normal University (Philosophy and Social Sciences Edition)*, 2020, No. 1.

He Haiwu, Yan An, Chen Zehua. Survey of Smart Contract Technology and Application Based on Blockchain. *Journal of Computer Research and Development*, 2018, No. 11.

Wang Han. Research on the Development Trend of Blockchain-Based Social Welfare Industry. *Technology and Economic Guide*, 2018, No. 36.

Gao Hang, Yu Xuemai, Wang Mao Lu. *Blockchain and New Economy: Era of Digital Currency 2.0*. Publishing House of Electronic Industry, 2016.

Hestenes D. "Modeling games in the newtonian world". *Am. J. Phys*, 1992.

Zhu Hong. The Shift of Interpersonal Trust: An Empirical Study of Interpersonal Trust. *Academia Bimestrie*, 2011, No. 4.

Zhang Hua. Digital Existence Communities and Moral Transcendence. *Morality and Civilization*, 2008, No. 6.

Yan Hui. *Social Classes in China's Digital Society*. National Library of China Publishing House, 2013.

Inspur, International Data Corporation. 2019 Report on Data and Storage Development Research. Inspur. 2019. https://www.inspur.com/lcjtww/2315499/2315503/2315607/2482232/index.html.

Chang Jia et al. *Blockchain: from Digital Currency to Credit Society*. China CITIC Press, 2016.

Jin Jianbin. The Construction of Social Trust in the Network Age: An Analytical Framework. *Theory Monthly*, 2010, No. 6.

Xiong Jiankun. The Rise of Blockchain Technology and the New Revolution of Governance. *Journal of Harbin Institute of Technology (Social Sciences Edition)*, 2018, No. 5.

Wang Jianmin. The Maintenance of Relationships in Chinese Society in the Transitional Period: From 'Acquaintance Trust' to 'Institutional Trust'. *Gansu Social Sciences*, 2005, No. 6.

Liu Jianming. "Big Data" is Not Omnipotent. *Beijing Daily*, May 6, 2013, p.18.

Zhang Jingyu, Li Zhihong. The Analysis of Alienation of Digital Identity. *Studies in Dialectics of Nature*, 2018, No. 9.

Wang Junsheng et al. Study on the Application of Digital Identity Chain System. *Power Communication Technology Research and Applications*, 2019, No. 5.

Jingdiwangtian et al. *Blockchain World*. CITIC Press, 2016.

Zhao Lei. Operating Mechanism of Block Chain and Its Regulatory Logic. *The Chinese Banker*, 2018, No. 5.

Wang Maolu, Lu Jingyi. Blockchain Technology and Its Application in Government Governance. *E-Government*, 2018, No. 2.

Xu Mingxing et al. *Blockchain: Reshape the Economy and the World*. China CITIC Press, 2016.

Lv Naiji. A Layman's View of Blockchain. *Science and Technology of China*, 2017, No. 1.

Project team of Yibin Central Sub-branch of the People's Bank of China. Digital Currency Development Application and Monetary System Change Discussion—Based on Blockchain Technology. *Southwest Finance*. 2016, No. 5.

Liu Qianren et al. Application and Research of Digital Identity Based on Blockchain. *Designing Techniques of Posts and Telecommunications*, 2019, No. 4.

Qiu Renzong et al. Ethical Issues in Big Data Technology. *Science and Society*, 2014, No. 1.

Liu Ruofei. China's Blockchain Market Development and Regional Layout. *China Industrial Review*. 2016, No. 12.

[US] Jihong W. Sanderson. *The Era of Everyone: Solutions for the New E'economy*. CITIC Press Group, 2015.

[UK] Peter B. Scott-Morgan. *How Five Unstoppable High-Tech Trends Will Dominate Our Lives and Transform Our World*, translated by Wang Feifei. China Machine Press, 2017.

Li Sheng. "Digital Divide": A New Perspective for Analyzing Modern Social Stratification. *Chinese Journal of Sociology*, 2006, No. 6.

Chen Shiwei. Discussion on Ethical Governance of Digital Divide in Big Data Era. *Innovation*, 2018, No. 3, pp. 20-21.

[Poland] Piotr Sztompka. *Trust: A Sociological Theory*, translated by Cheng Shengli. Zhonghua Book Company, 2005.

Que Tianshu, Fang Biao. Blockchain Technology Reshapes Social Consensus in the Age of Intelligence. *Chinese Social Science*, October 23, 2019, p. 5.

Wen Tingxiao, Liu Xuan. The New Order Theory of David Weinberger and Its Enlightenment of Knowledge Organization. *Library Journal*, 2013, No.3.

[France] Tocqueville. *De la démocratie en Amérique (Vol.2)*, translated by Dong Guoliang. The Commercial Press, 1990.

[US] David Weinberger. *Everything Is Miscellaneous: The Power of the New Digital Disorder*, translated by Li Yanming. Shanxi People's Publishing House, 2017.

Wang Wenqing. Modeling Theory in Science Education. *Science and Technology Information*, 2011, No. 3.

Qian Xiaoping. Reflections on a Few Issues on the Issuance of Digital Currency in China. *Business and Economy*, 2016, No. 3. p. 23.

Mi Xiaowen. The Impact and Countermeasures of Digital Currencies on Central Banks. *South China Finance*, 2016, No. 3.

Han Xuan, Liu Yamin. Research on Consensus Mechanism of Blockchain Technology. *Netinfo Security*, 2017, No. 9.

Han Xuan, Yuan Yong, Wang Feiyue. Security Problems on Blockchain: The State of the Art and Future Trends. *Acta Automatica Sinica*, 2019, No. 1.

Qiu Xun. China's Central Bank Issuing Digital Currency: Paths, Problems and Coping Strategies. *Southwest Finance*, 2017, No. 3.

Li Yi. Four Characteristics of the Operating State of "Digital Society". *Study Times*, August 2, 2019, p. 8.

Zhang Yihui, Wei Kai. Blockchain Reshapes Digital Identity; What Applications to Expect? *People's Posts and Telecommunications News*, April 11, 2019, p.7.

Liu Yizhong, et al. Overview on Blockchain Consensus Mechanisms. *Journal of Cryptologic Research*, 2019, No. 4.

Jin Yongsheng. Grasping the Essence and Growth Model of "Internet+". *People's Daily*, September 21, 2015, p.7.

Xu Zhong, Zou Chuanwei. What Can Blockchain Do and Cannot Do? *Journal of Financial Research*, 2018, No. 11.

Guo Zhonghua, Liu Xunlian. *Citizenship and Social Class*. Jiangsu People's Publishing House, 2017.

Chapter 4
Theory of Smart Contract

Since no man has a natural authority over his fellow men, and force creates no right, we must conclude that conventions form the basis of all legitimate authority among men.

−Rousseau, French Enlightenment thinker

Trust is the lubricant of economic exchange, the effective mechanism for controlling contracts, the implicit contract, and the unique commodity that is least likely to be bought.

—Kenneth Joseph Arrow, Nobel prize-winner in economics

Many of the rules, if programmed and executed on the blockchain currently constructed, might make the entire society better ordered and more efficient.

—Chen Zhong, professor of Peking University

4.1 Trust Digital Economy

More than two hundred years ago, in *The Wealth of Nations*, Adam Smith used the term "invisible hand" to describe the role of market mechanism in the operation of economy. Nowadays, along with the convergence and integration of big data, artificial intelligence, Internet of Things (IOT), Blockchain, 5G, and other digital technologies with human production and life, the productivity and production relations that drive human society forward are constantly changing. The world is now in an era of great changes in the accelerated transition from an industrial economy to a digital economy. The core factor of production in agricultural economy is land while that in industrial economy is technology and capital, but the core factor of production in digital economy is data. The blockchain-based Smart Contract system could make data rights effectively confirmed through the data uploading of "trust digitization", so the authenticity and security of data transactions and sharing can be guaranteed. This system also allows trust to flow as freely as information, thus building an efficient, authentic, transparent, and equivalent trust in digital economic ecosystem. There is no doubt that the perfect combination of blockchain technology and Smart Contract will greatly promote social changes and reconstruct existing productivity, means of production, production relations, and even rules and orders of human society. The

L. Yuming, *Sovereignty Blockchain 1.0*,
https://doi.org/10.1007/978-981-16-0757-8_4

combination will also provide a unified consensus mechanism for trust interactions among digital objects, thus helping form a credible digital economy in the future. Then a new contractual society will naturally emerge in which everyone works for the public, does his part, and finds his niche.

4.1.1 Blockchain Empowers Financial Technology (FinTech)

Finance is the heart of modern economy and the blood of real economy, and technological progress has been the main drive of the development and reform of financial industry. In recent years, with a new generation of information technology, financial industry has become increasingly integrated with information technology due to its high dependence on information data, which has accelerated financial innovation. In March 2016, the Financial Stability Board (FSB), core body for global financial governance, defined "FinTech" for the first time as follows. FinTech, financial innovations brought by technologies, can create new business models, applications, processes, or products, and thus have a significant impact on financial markets, financial institutions or the way financial services are provided (Jianqing 2017). According to the above concept, the key to FinTech is the interaction and integration of finance and technology, and technological breakthrough is the driving force of FinTech. Combined with the role of information technology in driving finance, FinTech can be divided into three stages (Table 4.1). As can be seen, FinTech is thriving in the context of a new round of technological revolution and industrial changes. The deep integration of big data, artificial intelligence, blockchain, and other new-generation information technologies with financial services has become the major driving force of financial innovation.

In September 2019, the People's Bank of China (PBOC) issued the *FinTech Development Plan (2019–2021)* (hereinafter referred to as the *Plan*), which establishes the

Table 4.1 Main stages of technology-driven fintech

FinTech development stage	Time	Driving technology	Main forms	Universal access	Relationship between technology and finance
First stage	2005–2010	Computer	ATM, Electronic bill	Lower	Technology as a tool
Second stage	2011–2015	Internet	Third-party payment, P2P network lending	High	Technology drives changes
Third stage	Since 2016	Big data, Blockchain, Artificial intelligence, etc.	Smart finance	Higher	Technology and finance are deeply integrated

top-level plan for China's FinTech. This *Plan* is conducive to establishing a more comprehensive FinTech system, putting the current disorder of industrial development to an end, effectively avoiding the waste of FinTech resources, and encouraging FinTech entrepreneurs to develop financial technologies with unique scenarios. Meanwhile, the *Plan* also points out that, "with the help of machine learning, data mining, Smart Contract and other technologies, FinTech can simplify the transactions between the supply and demand sides and reduce the marginal cost of financing". The plan also emphasizes the need to explore the application of emerging technologies in optimizing a trusted environment for financial transactions, and steadily promote the pilot test and application of technologies such as distributed ledgers. Blockchain is an emerging distributed ledger technology. Its Smart Contract is highly secure and tamper-resistant digital protocols that can be deployed on distributed ledgers to provide guaranteed execution and processing in a trust-free manner. The combination of blockchain and Smart Contract can be applied to digital currency, digital ticket, securities trading, financial audit and many other financial fields.

Blockchain: Underlying Core Technology of FinTech. The FinTech that drives financial development has shifted from the application level such as mobile internet, big data, and cloud computing to the underlying technology innovation such as Blockchain. Blockchain is a decentralized ledger system that can make up for the deficiencies of traditional financial institutions, improve operational efficiency, reduce operating costs, update market rules flexibly, prevent information tampering and forgery, and also enhance stability. Blockchain, an infrastructure of trust links, can deal with the trust of financial activities at a low cost. Blockchain also evolves financial trust from bilateral trust or the establishment of a central trust mechanism into multilateral trust and social trust, seeking ways to find credibility with common trust.[1] In addition, blockchain, conducive to regulation, becomes a part of regulatory technology, facilitating regulators' access to more comprehensive and real-time regulatory data. It can be said that blockchain technology will be the key underlying infrastructure of Internet finance and even the entire financial industry. As the underlying technical architecture of FinTech, blockchain is bound to reshape financial forms in many aspects. Whether in traditional financial services, Internet financial innovations such as crowdfunding and P2P network lending, or in preventing financial risks, strengthening financial supervision, and combating illegal fund-raising and other fields, blockchain technology has very broad prospects for application. Internet finance is entering the era of "Blockchain Plus". In short, as the basic technology of financial innovation, blockchain has great strategic value and has been widely recognized by central banks and financial institutions of many nations, who are studying its applications in deepening financial reforms, enhancing financial supply, promoting financial innovations, improving financial credit, and preventing financial risks. Blockchain technology is bound to be more widely used in the financial sector as well as economic, social, and other fields.

---

[1] Huo Xuewen. Blockchain Will Become the Underlying Technology of Fintech. NetEase. 2016. http://tech.163.com/16/0710/10/BRJT0K4400097U7R.html.

Blockchain: the Important Infrastructure for Digital Finance. From the financial perspective, blockchain and digital currency are the new generation of digital financial system. "The blockchain digital finance built by the basic laws of the digital world and mature technology is extremely aggressive. BTC (Bitcoin) has built the industrial cornerstone of digital finance in just a decade" (Jiwei 2019). In the context of financial digitization, it has become a global consensus that digital currency has advantages such as facilitating commercial activities, reducing issuing costs, and improving the capacity of central banks to control currency circulation. "Digital finance may restructure financial operation mode, service model and even the entire ecosystem. Concise, free and open, it transcends temporal and physical boundaries, breaks national borders, and respects the autonomy and willingness of market participants."[2] Driven by FinTech, digital finance does not need to rely on traditional financial intermediaries, and enables assets to flow on the premise of maintaining the original full amount of information. Traditional financial services will be logically encoded into transparent, trustworthy, self-executing, and enforceable Smart Contract. Smart Contract provides a variety of financial services, and a Smart Contract even represents a financial form. In this sense, controlling Smart Contract means controlling future financial business. Therefore, on the basis of secure and efficient user authentication and authority management, a Smart Contract must be verified by relevant authorities before being chained, so as to ensure whether its program can run in accordance with the expected policies of regulatory authorities. When necessary, regulators can prevent ineligible Smart Contract from being chained or shut down local residents' rights to execute Smart Contract, and establish licensing regulatory intervention mechanisms that allow code suspension or termination. The application of blockchain has expanded from the initial finance to smart manufacturing, Internet of Things, supply chain management, data backup and transactions, and many other fields. In the future, blockchain will definitely change current social business model, create a new digital financial system, and then trigger a new round of technological innovation and industrial reform.

Blockchain: a Powerful Engine of Financial Inclusion. "Financial inclusion aims to provide appropriate and effective financial services at an affordable cost to all segments of society and groups in need of financial services, based on the principles of equal opportunity and business sustainability".[3] In the development of financial inclusion, there are certain difficulties in "universal access", "benefit", and "financial sustainability". The point-to-point network model of blockchain is highly autonomous and open, enabling financial services to gradually penetrate into remote areas and provide financial services to relatively vulnerable groups. It can break traditional geospatial constraints, provide financial services for groups that have long been financially excluded, and improve the universality of financial services. On the basis of ensuring information authenticity and validity, information digitalization,

---

[2] Yao Qian. Crypto-Currencies Have the Potential to Become Real Currencies. The Paper. 2019. https://www.thepaper.cn/newsDetail_forward_4445573.

[3] The State Council of the People's Republic of China. The 2016-2020 Plan on the Development of Inclusive Finance (GF [2015] No.74). 2015.

assets and work processes can be realized through embedding blockchain and Smart Contract. Besides, groups excluded from the financial system due to information asymmetry can be included in the financial system. Through information sharing in less developed regions, monopolies can be avoided, and coverage of financial services can be expanded. Reducing labor costs and increasing efficiency also empower financial institutions to provide financial services at lower prices to benefit more people. In addition, in terms of targeted poverty alleviation and other aspects, the whole process of data monitoring through blockchain can make participants highly transparent and social resources completely controllable. As it is difficult to tamper with the flow of funds on the blockchain, it is easy for the authorities to effectively manage targeted funds and address integrity.

Blockchain: a Strong Support to Prevent and Defuse Financial Risks. The essence of finance is risk, and the essence of risk is information asymmetry. Blockchain can tackle information asymmetry and difficulty in trust between financial industry and real economy. This will greatly reduce the cost of the financial industry and help prevent financial risks, while Smart Contract is effective in improving the efficiency of financial services for real economy. In the blockchain financial application scenario, technical trust can replace commercial credit to a certain extent within certain limits, but it does not negate traditional ways of trust. In essence, it is not about de-trust, but about empowering commercial credit with technical trust, which is conducive to maintaining the principle of contract and financial integrity.[4] Blockchain is a machine that builds trust, not only among strangers through distributed ledgers and consensus algorithms, but between financial institutions and MSMEs (micro-, small-, and medium-sized enterprises). Through the supply chain finance platform jointly established by the government, enterprises, and financial institutions, blockchain helps realize the authenticity, credibility, and sharing of business data of MSMEs, greatly reducing costs of banks' due diligence and credit risks with the authorization of enterprises. Smart Contract can also facilitate smart lending, smart interest collection, and smart risk control, so as to help finance better serve real economy and solve difficulties and high cost of financing for MSMEs (Yongxin 2019). In addition, the parameter setting of Smart Contract is a means of regulation. Just as regulatory indicators such as statutory deposit reserve ratio and capital adequacy ratio that are used to prevent and control bank risks, regulatory authorities can also adjust or intervene in the parameters of Smart Contract to control the scale and risks of financial services. For example, the application of blockchain technology in bill business imposes specific restrictions on commercial agreements by using programmable Smart Contract, and introduces monitoring nodes to confirm transaction parties, thus ensuring the uniqueness of value exchange and effectively defusing financial risks.

---

[4]Li Lihui. Five Steps to Build Digital Social Trust Mechanism. finance sina. 2018. https://finance. sina.com.cn/hy/hyjz/2018-12-15/doc-ihmutuec9368548.shtml.

4.1.2 Smart Contract and Data Transaction

According to the latest report from International Data Corporation (IDC), China will generate and replicate data at an annual rate 3% higher than global average, which produces about 7.6 zettabytes (ZB) of data in 2018, and will produce 48.6 zettabytes (ZB) by 2025 (Reinsel et al. 2019). Data is a key factor of production in digital economy. Data transactions drive data flow, release data value, improve big data industry chain, accelerate transformation and upgrade real economy to digitalization, so data becomes a powerful engine for digital economy. As the first comprehensive big data pilot zone in China, Guizhou Province takes the lead and sets a model nationwide in actively exploring data source opening, data authentication, data pricing, data standards, data right confirmation, data security, and other fields related to data transactions. In April 2015, Global Big Data Exchange (GBDEx), the first big data exchange in China, was officially put into operation. Its establishment was hailed by IDC as "an important milestone in the development of global big data industry, opening a new chapter in the history of big data transactions". Digital economy takes data as energy. The data transaction located in the upstream of big data industry chain can drive the circulation of data elements and become a key field for the development and application of blockchain technology.

At present, due to the unclear ownership of data, massive data resources are subject to government and enterprise data barriers and data isolation, which restricts the flow of data elements, as well as the development of big data industry and digital economy. There exist problems, e.g., unclear definition and ownership of transaction data, inability of important data traceability, and personal privacy disclosure. If these problems cannot be effectively solved, they will seriously infringe upon people's legitimate rights and interests and hinder the sustainable development of big data industry. In this case, how to effectively use blockchain to realize big data transaction has become an urgent problem. Blockchain can help participating entities to build trust with each other and facilitate data transactions. Data ownership, transactions, and scope of authorization are recorded on the blockchain, where data ownership can be confirmed. A refined scope of authorization can regulate the use of data (Huawei Cloud 2018). In 2017, Global Big Data Exchange formulated the *Big Data Transaction Blockchain Technology Application Standard*, and applied the blockchain technology to trading system for trusted transaction of data assets, enabling data rights confirmation and data sources tracing. The addition of blockchain technology allows each step of data transaction to leave a time stamp and accurately record data generation, exchange, transfer, updating, development, and utilization. It is convenient to confirm, trace, manage, and access transaction data, realizing data security and privacy protection. Therefore, the secure flow of data has been accelerated due to the perfection of rules. Blockchain of big data transactions promotes the streamlining, transparency, and standardization of transactions, facilitate data confirmation and quality assessment, and provide a solid safeguard for big data transactions (Key Laboratory of Big Data Strategy 2017).

Blockchain Cracks Data Sharing and Security Challenges. Data transaction is in fact paid data sharing, and sharing and security is contradictory. Blockchain technology, however, reconciles this contradiction. First of all, blockchain technology protects data from being leaked via multiple encryptions, and the Enigma system based on blockchain technology can operate data without accessing original data, which effectively protects the privacy of data and provides a solution to data security in the context of big data transactions, especially privacy protection. Second, based on multiple encryptions and combined with digital signature technology, blockchain can ensure data is only available to authorized personnel. Meanwhile, if the data is shared among all nodes, each node will have a copy of encrypted data that can only be decrypted using the corresponding private key. This technology not only achieves the selective sharing of data, but guarantees data security. "Blockchain distributed architecture is more suitable for creating a trusted data-shared ledger among multiple stakeholders. It allows multiple stakeholders to realize data interaction and sharing in a polycentric manner without a 'centralized' authority".[5] The data transaction scheme based on blockchain Smart Contract records the buying and selling behaviors of both parties through blockchain technology, and realizes automatic transactions with Smart Contract. The transaction data is not recorded in the blockchain. The data will be encrypted and stored in an external data memorizer which will be available to the buyer when the Smart Contract takes effect. The Smart Contract is executed on multiple nodes, and the executed results must be the same. The results produced by Smart Contract must reach a consensus before they can be accepted.[6] In addition, the buyer can also release transactions through the regulatory blockchain consensus network to achieve third-party witness, which also increases the security of data transactions.

Blockchain Guarantees the Legitimate Rights and Interests of Parties in Data Transactions. The data ownership in data transactions is highly controversial, and there is ambiguity as to who the data rights holders are. Data is easily replicated over the network, and ownership protection is difficult to implement. The use of blockchain technology to authenticate data can clarify the source, ownership, right to use, and circulation path of the data, so that the transaction record is recognized by the whole network, transparent and traceable, thus protecting the legitimate rights and interests of both parties in data transactions. On the one hand, blockchain provides a traceable path that can effectively crack falsified data. Blockchain completes the calculation and recording of data through multiple nodes participating in computing in the network, and verifies the validity of data among each other, which is conducive to the anti-counterfeiting of transaction data and guarantee of data users' legitimate rights and interests. On the other hand, blockchain erodes the threat of intermediary

---

[5]Zhang Yuwen, Chi Cheng, Li Yurong. Can Blockchain Break the Dilemma of Data Interaction? The China Academy of Information and Communications Technology. 2019. http://www.caict.ac.cn/kxyj/caictgd/201912/t20191212_271577.htm.

[6]Cai Weide, Jiang Jiaying. Three Important Principles of Smart Contract. Sohu. 2019. http://www.sohu.com/a/290611143_100029692.

centers copying data and helps establish a reliable environment for big data transactions. The intermediary centers have the condition and capacity to copy and preserve all the data flowing through. However, the blockchain based on decentralization can eliminate the threat of intermediary centers copying data, and protect the legitimate rights and interests of data providers. Smart Contract enable data to be effectively confirmed. Through the data linking of "trust digitization", the authenticity of data can be effectively guaranteed and the problems that used to be hard nuts can be solved for the industry. Besides, various industries can be transformed and upgraded from many aspects, such as "cost reduction" and "efficiency improvement".

Blockchain Reduces Security Risks in Data Transaction Systems. Hackers and server outages are the biggest concerns for data transaction systems. Once the data transaction system is damaged, the consequences are incalculable. The distributed network structure of blockchain enables the whole data transaction system to work without centralized hardware or organization. When a node is damaged, it will not cause the loss of data on other nodes. The whole data transaction system can operate as usual, and blockchain big data transaction will proceed smoothly. In addition, with the help of blockchain technology, any participating node can verify the authenticity and integrity of the ledger content and the transaction history constructed by the ledger. This ensures that the transaction history is reliable and tamper-resistant, improves system traceability, and reduces the cost of trust in data transaction systems. The data transaction scheme based on the blockchain Smart Contract is that all nodes between the transaction consensus network and the regulatory consensus network jointly maintain a ledger record through blockchain. Data transactions can be released through all nodes. The mentioned data transaction Consensus Network is released to the blockchain in the form of Smart Contract, and distributed to each node through peer-to-peer network to ultimately reach consensus. Consensus is also reached by releasing data transactions to the regulatory Consensus Network, and Smart Contract can be automatically executed through the verification nodes of the data transaction Consensus Network. By linking data fingerprints and combining them with digital signature technology to ensure the authenticity and integrity of the data, authorized certification bodies can verify the data and link authentication results, thus achieving credit enhancement of data. Blockchain technology helps solve end-to-end value delivery of trust, creates trust connections for more participants, and provides trust data sharing services in a low-cost, efficient, transparent, and peer-to-peer manner.[7]

Blockchain Ensures the Security and Credibility of Data Transaction. Based on blockchain technology, a big data transaction platform adopting decentralization system architecture and computing paradigm can optimize big data transaction by achieving security, controllability, and comprehensive supervision of data asset registration, data value assessment, data asset preservation, data asset investment, data transaction settlement, and other steps in the transaction process, thus facilitating the

---

[7]Zhang Yuwen, Chi Cheng, Li Yurong. Can Blockchain Break the Dilemma of Data Interaction? The China Academy of Information and Communications Technology. 2019. http://www.caict.ac.cn/kxyj/caictgd/201912/t20191212_271577.htm.

flow and application of data. Blockchain is conducive to a data assets right registration platform with "Credibility", avoiding disputes over data assets ownership, and promoting the orderly process of transaction confirmation, account reconciliation, and investment settlement. The traceable and tamper-resistance of Smart Contract is conducive to a complete flow of big data transactions. By comparing the same category of data and transaction history on blockchain, the newly registered data assets can be evaluated reasonably. Combining traditional means of assets preservation with blockchain technology effectively protects the integrity of data assets and prevent data assets from being attacked, leaked, stolen, tampered and illegally used. In data assets investment, blockchain technology is applied to the registration, evaluation, preservation, investment and other steps of data assets to ensure the credibility and security of data assets ledgers. Blockchain Smart Contract technology can greatly simplify the transaction settlement services and transaction capital account management. Meanwhile, Smart Contract can write complex rules of data transaction settlement into computer programs in the form of contract terms. When an action that satisfies the contract terms occurs, subsequent actions such as receiving, storing, and sending will be automatically triggered, making data transaction settlements intelligent.

4.1.3 Trusted Digital Economy Ecosystem

Human production and life as well as a series of production factors are being digitized, and digital economy has become a new form of socio-economic development (Chongming 2019).According to the data from China Academy of Information and Communications Technology (CAICT), China's digital economy aggregate in 2018 reached 31.3 trillion yuan, accounting for 34.8% of GDP with an increase of 1.9 percentage points compared to last year. Digital economy is a new economic form with digital knowledge and information as the key factors of production. With digital technology innovation as the core driving force and modern information network as the important carrier, digital economy will steadily improve the digital and intelligent level of traditional industries, and accelerate the restructuring of economic development and governments' governance mode through the deep integration of digital technology and real economy (China Academy of Information and Communications Technology 2019). Digital economy is a more advanced economic stage after agricultural economy and industrial economy. To understand the digital economy, we need to shake off the inherent thinking mode of economy, broaden horizon and regard it as a leap-frog development of industrial economy and agricultural economy, which expands the space and practical dimension of traditional economic model, though it has similar economic paradigm and principles with industrial economy and agricultural economy (Hongli and Zhongyi 2019). In the era of agricultural economy, production factors are land and labor. In the time of industrial economy, the importance of land has declined, and productive capital (such as machinery and equipment) and labor are regarded as two major production factors, while the implicit assumption is that land is included in productive capital. In the era of digital economy, in addition to capital and labor, data has become another core production factor.

As a new means of production, data has become the core production factor and new infrastructure to drive economic growth. The information Internet model based on centralized system is no longer suitable for the rapid development of digital economy. As a technology system that is able to realize distributed storage, tamper-resistance, and prevent repudiation, blockchain creates a peer-to-peer value transmission network. With the help of cryptography and other technologies, blockchain realizes data value confirmation and assets digitization, and ensures the security and reliability of circulations such as data sharing. It is characterized by security, trust, clear digital property rights, co-governance, sharing, etc., and provides a secure and credible environment for digital economy (Fig. 4.1). Therefore, the consensus mechanism and incentive mechanism of blockchain can replicate the organizational structure under real economic system, improve the efficiency of value delivery, reduce costs, and become the infrastructure and important component of constructing digital economy and digital society.

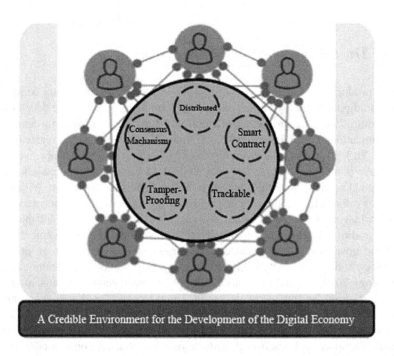

Fig. 4.1 Blockchain: the machine of trust. *Source* The Information Office of the People's Government of Guiyang Municipality: *Blockchain Development and Application of Guiyang*, Guizhou People Press, 2016

Blockchain Makes it Easier to Confirm and Partition Data. Data confirmation is the fundamental requirement and legal basis for the sound and secure development of digital economy. Data confirmation refers to the process of dividing and confirming all the rights and interests in data rights set, including ownership, right of possession, right of dominion, and right of privacy. Data confirmation is essentially a solution to the problem that data is unlabeled and easy to be copied and abused.[8] The collection, storage, use, circulation, and destruction of data will give rise to multiple ownership relationships, which are difficult to define clearly. Data partitioning is ancillary to data confirmation. The root of this problem is the difficulty of data partitioning and data flow due to the inability to confirm data (Hongli and Zhongyi 2019). The definition of data ownership is the basic support and guarantee for digital economy. If the ownership of data in various relationships cannot be clearly defined, the orderly circulation of data cannot be achieved. The difficulty of data partitioning will endanger the multi-dimensional and refined development of digital economy, and fail to provide reliable rights protection for new formats and models, thus worsening digital economy. The distributed trust system established by blockchain technology provides a basic architecture guarantee for data confirmation and partitioning. Its natural decentralized ledger property ensures the authenticity and tamper-resistance of data, and provides a cornerstone for the data-driving era of digital economy.

Blockchain Creates a Credible and Secure Digital Economy. The digital economy built by blockchain is an economic system based on "technology trust". In the Internet model, the centralized storage of mass data collects, exchanges, and flows information at a low cost, but it fails to guarantee the security and credibility of digital information in digital economic activities. The Consensus Mechanism and Smart Contract of blockchain construct the rule protocol of data generation, transmission, calculation, and storage in its centralized environment, and support the secure flow of value with data, information, and knowledge as carriers. In addition, with cryptography algorithm, the encryption and protection of the data on blockchain is gradually turning mature, so as to achieve the "end-to-end" privacy protection in network nodes. From this perspective, blockchain has realized the basic protocol of the Internet of Value, and becomes a strategic support technology for digital economy, which helps establish secure and credible digital economic rules and order. Blockchain uses encryption and consensus algorithms to establish a trust mechanism, thus making repudiation, tampering, and fraud costly. Meanwhile, it ensures that data is not tampered or forged to achieve integrity, authenticity, and consistency of data. A credible and secure digital economy based on blockchain creates a more credible

---

[8]From the semantic understanding, data confirmation is to confirm the right holders of data. The rights include the right of ownership, right of use and right of usufruct. From the perspective of business, data confirmation is the process of clarifying the rights, responsibilities and relationships of the data trading parties in the commercial process, so as to protect their legitimate rights and interests. (Liu Quan. *Blockchain and Artificial Intelligence: Building an Intelligent Digital Economy World*. Posts and Telecom Press, 2019, p. 42).

and secure market environment for sovereignty economies to participate in global digital economy, and promotes closer digital economy cooperation between nations, regions, businesses, and individuals. The operation of economy and society will be more transparent, thanks to the "information on-chain" of the government, enterprises, and individuals, and the information flow between various subjects is smooth and credible.

Blockchain is the "Key" to Assets Data. Asset datafication is a necessary means to map physical assets (tangible assets and intangible assets) to data assets and enrich the connotation of the latter. It is also the only way to expand the space and dimension of digital economy and drive digital economy toward digital finance and digital society. Data asset, a new form of asset under digital economy, mainly includes three aspects: first, legal digital currency and other digital tokens, such as Bitcoin and Litecoin; second, digital financial assets, including digital stocks, crowdfunding equity, private equity, bonds, hedge funds, and other types of financial derivatives, such as futures, options, and other financial assets; third, all kinds of assets that can be digitalized. In practice, Smart Contract can be run on the blockchain by giving data assets the form of code and then external contracts can be used to trigger the automatic execution of the contract, thereby determining the reallocation or transfer of data assets in the network. Its contract mark can be material property rights such as cars and houses or immaterial property rights such as stock equity, bills, and digital currency. It can be seen that blockchain technology makes it easier to confirm, partition, and share data. The distributed network topology of blockchain can expand the application of digital assets and take the lead in the application of digital currency, data transaction and other economic fields to prevent the replication of data assets.

Blockchain is a Major Component of a Trust System in Digital Society. Digital economy is the premise, foundation, and core driving force of the[9] progress of the digital society. And digital society is the ultimate goal of digital economy, which provides cultural-ethical power, intellectual support, and necessary conditions for digital economy. Compared with traditional industries, digital economy responds faster to the market, with lower investment thresholds, simpler production links, and lower costs. Therefore, digital economy can not only achieve low input with high yield, reduce costs, and improve efficiency, but contribute to the sustainable development of economy and society as well. It can be said that digital economy will become a major driving force for social productivity upgrading and a new engine for economic growth. Decentralized digital currency issuance, tamper-resistant contract terms, and digitization of asset transactions of Smart Contract based on blockchain technology make various restrictions of any contract simpler, with lower transaction costs and higher transaction efficiency. All information on blockchain is recorded in

---

[9]The rapid development and wide application of digitization, networking, big data, artificial intelligence and other contemporary information technologies have given birth to the specific technological, social construction and social cultural form— "digital society". The particular reference to "digital society" is a more visual expression of "cyber society" or "virtual society". (Li Yi. Four Characteristics of the Operation State of "Digital Society". *Study Times*, August 2, 2019, p.8).

digital forms, which means workforce will be significantly reduced and each piece of information can be recorded, confirmed, and authenticated on blockchain. The digital economy under blockchain model will reconstruct the relationship between the means of production and laborers. Firstly, blockchain relies on cryptography technology, tamper-resistant distributed ledgers, and peer-to-peer networks to establish a right confirmation mechanism that defines ownership and rights of use between the means of production and laborers; second, blockchain completely replicates offline economic relations in digital society through Consensus Mechanism, permission control, and other technologies; third, the integral mechanism and incentive mechanism in blockchain will redefine the distribution rules of means of production, enrich and stimulate the innovative models and ideas of digital economy (Hongli and Zhongyi 2019). By realizing peer-to-peer connection between people, machines, and networks, blockchain enables large-scale connections to break the barriers between humans and human, between humans and objects, and to accurately exchange needed goods and services. It can truly connect everything, release huge energy in digital economy, and lead us into a zero marginal cost society. In short, the combination of blockchain technology and Smart Contract will transform social productivity and production relations, change the way of human production and life, and build a major component of trust digital society (Fig. 4.2).

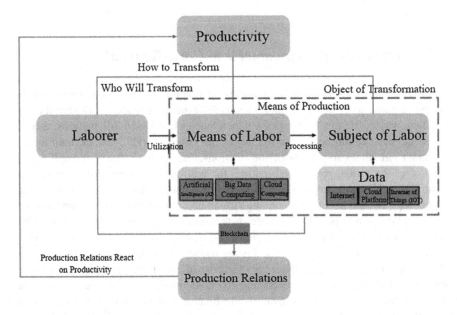

Fig. 4.2 The structure of social relations in digital society under blockchain

4.2 Programmable Society

In *The Theory of Moral Sentiments*, from the perspective of altruism and the basic principle of compassion, Adam Smith revealed the basis on which human society relies for its maintenance and harmonious development, as well as the general moral standards that human behavior should follow. However, as he discussed in *The Wealth of Nations*, "everyone pursues his own interests", which is a human nature and a natural phenomenon. As a machine to construct trust, the core value of blockchain is the "technology contract" to offer trust for human society, so as to provide a new solution to the reconciliation and harmony between human and human, and between humans and nature. Human civilization has evolved from a "Status Society" to a "Contractual society". Smart Contract can replace all paper contracts, integrate the physical and virtual worlds, and ensure credible interactions with real-world assets. In the future, most human activities will be completed on blockchain, thus realizing the programmability of trust, asset, and value, which will completely change the way of value delivery in the human society and help form a programmable world. Meanwhile, digital terminals such as digital citizens and smart devices will be interconnected on the blockchain system. Driven by multiple factors such as data, algorithms, and scenarios, digital terminals will build a brand-new and absolutely reliable social contract relationship that does not require a third-party endorsement, and innovate social governance sustainably to lead humans to a more efficient, fair, and orderly smart society.

4.2.1 From Being Programmable to Cross-Terminal

In the real world, everyone is in a variety of relationship contracts, under which everyone participates in the production and life of the whole society. "Digital society is a new way of life that transfers the dependencies in people's life patterns, credit, laws, and even culture in the real society to the virtual world".[10] Blockchain Smart Contract virtualizes social production relations, enables the value of the real world to flow in the virtual world, and realizes the consensus modeling of the different contracts and business processes of the real world in the virtual world. If blockchain technology is ultimately to promote the virtualization of production relations and the development of productivity, the core of the entire blockchain ecosystem must be able to support various contracts, namely, business contracts, and share transaction books among relevant participants. Just as money is the foundation of finance in the real society, money and finance are the core of social operation. Programmable money is the core and value exchange foundation of programmable finance and programmable society, and programmable finance is the center of programmable society. At present, the programmable freedom of blockchain is limited, but it can

---

[10]Hu Kai. The Foundation of Digital Society—Smart Contract. CRI Online. 2018. http://it.cri.cn/20180531/ece4e6cc-c029-7ebd-7fda-c93d52e96f93.html.

be expected that with advancing technology, the programmable application range based on Bitcoin blockchain will become wider and wider. It can be said that Smart Contract and programmable society embody the ability to realize a highly automated, intelligent, fair, and law-abiding production relations of virtual society based on blockchain in the future. With the maturity and implementation of blockchain Smart Contract, profound changes will take place in economy, society, production, and life. We come to enter the era of "Artificial Intelligence & Internet of Everything" and programmable economic and social system.

From Programmable Currency to Programmable Economy. Currently, we are entering a new stage of informatization, that is, a stage of intelligence characterized by the deep mining and fusion of data. Under the background of three-way integration of people, machines, and things, with the goal of "everything needs to be connected and everything can be programmed", digitalization, networking, and intelligence present a new trend of integrated development (Hong 2018). "Programmable" means the completion of complex actions through pre-set commands, and the ability to judge external conditions and respond accordingly. Programmable money is the special use of some money in a specific time, which is of great significance for the government to manage special funds. The new digital payment system based on blockchain technology, with its decentralized, key-based, and unimpeded currency transaction model, ensures security while reducing transaction costs, which may have an enormous impact on traditional financial system. It also portrays an ideal transaction vision of global currency unification, which frees the dependence of currency on central banks for issuance and circulation. Programmable money brings programmable finance, which then brings programmable economy. Finance is an activity in which people exchange value across time and space, and the premise of value exchange is "information" and "trust". If programmable money is to decentralize money transactions, then programmable finance decentralizes financial market, which is the next key link in the progress of blockchain technology. New technologies are constantly penetrating into the complex dynamic process of economy, society, and life, bringing enormous changes to the operation mode of human society and its economic organization. However, all this will change with the advent of programmable economy. As a new economic model based on automation and digital algorithms, programmable economy writes the process of transaction execution into an automatic programming language and forces the pre-implanted instructions to run with code, to ensure the automaticity and integrity of transaction execution. It has brought us an unprecedented technological innovation. At the level of execution, it reduces the monitoring cost of transactions, and has bright prospects in reducing fraudulence, combating corruption, simplifying supply chain transactions and other "opportunistic behaviors". It guides the future of new economy. The scripting language of blockchain makes programmable economy a reality. The charm of script lies in its programmability as it can flexibly change the requirements of value retention to cater to social and economic activities, which is also an advantage of programmable economy. In this way, we can manage levels of economy, society, and governance through Smart Contract, which will play an important role in promoting the operation of economic system and society.

Blockchain and Programmable Society. Based on the programmable characteristic of blockchain Smart Contract and its application, programmable network society and economy that trust each other without third-party trust mechanism can be established. By tackling de-trust, blockchain technology provides a universal technology and a global solution, that is, improving the operating efficiency and overall level of the entire field without the help of third parties for establishing credit and sharing information resources. Blockchain 1.0 is the supporting platform of virtual currency. The core concept of Blockchain 2.0 is to take blockchain as a programmable distributed credit infrastructure to support Smart Contract. The application of blockchain extends from currency to other fields with contract functions, including house property contract, intellectual property right, equity, and certificate of indebtedness. Blockchain 3.0 not only extends its application to social governance, such as identity authentication, auditing, arbitration, and bidding, but to industrial, cultural, scientific, and artistic fields (Zhiming 2018). In this stage of application, blockchain technology will be used to connect people and devices to a global network, allocate global resources in a scientific way, realize the global flow of value, and promote the whole society to the era of smart interconnectivity. The consensus modeling of different fields and business processes in the virtual world will even create a contract service or process of new production relationship that integrates the real world and the virtual world, which can be described as programmable society. In this social background, just like humans, machines have intelligence and can communicate autonomously, forming laws and rules of programmable society automatically. They can iterate and evolve by themselves without help of external forces from humans. This is a popular, open, mutually beneficial, and smart organizational construction and deep collaboration platform for organizational units (individuals, groups, and institutions) which builds a borderless social ecology, facilitates organizational transformation and individual potential tapping, and leads value connection and collaborative innovation with human relationship as the core. From the historical perspective, changes within the boundaries of autonomous organizations will change not only the structure of their internal connections but the state of their external connections. Purposeful interaction behavior will promote the organization to spontaneously evolve into a higher form, thus changing the connection structure and regulation mode of the whole social organization.[11] Programmable society brings about trust reconstruction, reduces transaction costs, upgrades the efficiency of social management, and the way of social governance. It may eventually lead humans to a more just, orderly, and safe self-governing society.

Blockchain and "Artificial Intelligence & Internet of Everything". The programmability of blockchain will push society into an era of "Artificial Intelligence & Internet of Everything" based on machine trust. IDC predicts that by 2025, the number of IOT devices over the world will reach 41.6 billion, generating 79.4

---

[11] Nanhu Internet Finance Institute. The Programmability of Blockchain Technology Transforms the Organization Form of Economic Society and Life. Eastmoney. 2016.
    http://finance.eastmoney.com/news/1670,20160817656079298.html.

zettabytes (ZB) of data.[12] "The physical world was offline before the birth of Internet; after the advent of the Internet, the world is evolving into an online society" (Jian 2016). The connection is the result of being online, and it can penetrate into the society only after being online. Therefore, the essence of the Internet is "connection", making it possible for people to connect (smart Internet), things to connect (Internet of Things). and industries to connect (industrial Internet).[13] Blockchain technology subverts the lowest level protocol of Internet and integrates big data, artificial intelligence, Blockchain, 5G, and other technologies into Internet of Things. This integration will create a smartly connected world that has positive effects on all individuals, industries, societies, and economies, and propels human society toward a world where everything is perceived, connected, and intelligent. Intelligent connection based on ubiquitous network[14] requires the help of Internet, Internet of Things, and other connection technologies, as well as various intelligent terminal hardware facilities, so as to build a ubiquitous network world of "Artificial Intelligence & Internet of Everything". This connection will allow behavioral data to be available at anytime and anywhere, and timely delivered and stored over the network, which lays the foundation for computing and applications. The universal connection includes not only the digital connection between humans, but also the connection and transfixion among humans, objects, and smart terminals realized by digitization. Meanwhile, on the basis of universal connection, cross-domain connection further relies on the unique convenience of virtualization brought by digitization to break through space and geographical restrictions in a revolutionary way and realize the effective connection across regions, thus truly realizing the interconnectivity of global network integration. In the network world formed by cross-domain connection, any specific person, object, computer, intelligent device, or server, etc., exists as a "connection point" on the digital network (Yi 2019). "Each intelligent terminal is a microcomputer. It is a client that is connected to the server side via a wireless network. The data from the client and server side can be transferred each other. Obviously, the most significant feature of an intelligent terminal is connectivity."[15] This will allow more and more intelligent terminals of all shapes and sizes to be connected across PCs and mobile phones, and enable humans to better perceive and understand the world with the help of sensors. It is conceivable that every intelligent terminal worldwide could have a chip connected to the blockchain network in the future. Once this kind of blockchain IOT network is finalized, the interaction between human and machine, machine and machine intelligence will have a universal language, social rules will become programmable, and social resources can be freely connected.

---

[12]IDC. Global Internet of Things Device Data Report. asmag. 2019. http://security.asmag.com.cn/news/201906/99489.html.

[13]Fang Jiacai. *Linking: Interaction, Experiencing, Collaboration, Share and Win-win*. China Machine Press, 2015, p. 25.

[14]The concept of "ubiquitous networks", originated in the Latin word "ubiquitous", refers to ubiquitous networks.

[15]Wang Jiwei. Four of the Top 10 Signs of the Super-connected Era: Smart Terminal Momentum Has Come. Sohu. 2017. https://www.sohu.com/a/127131184_115856.

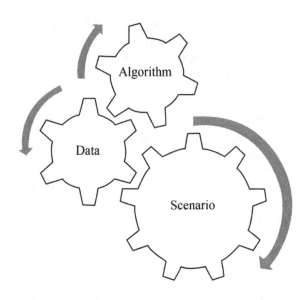

4.2.2 Data, Algorithm, and Scenario

"To a certain extent, Smart Contract is the digitalization of production relations, the purpose of artificial intelligence is the digitalization of things, and the purpose of Smart Contract is the digitalization of relationships between people, between humans and objects, and between things" (Biao 2019). Therefore, both artificial intelligence and Smart Contract are pointing to a smart society.[16] The mutual promotion and integration of artificial intelligence and Smart Contract technology not only bring more extensive application scenarios, data resources, and algorithm resources to Smart Contract, but make artificial intelligence more secure and efficient, so as to enhance the width and depth of blockchain field. The merging of the two technologies plays an essential role in removing trust intermediaries, reducing transaction costs, preventing risks, and so on. It also has a key impact on the construction of a smart society. Data, algorithm, and scenario are the three key elements of a smart society (Fig. 4.3). Among them, data is the basis, algorithm is the means, and scenario is the purpose. In the future, Smart Contract will also be equipped with predictive deductions based on unknown scenarios, computational experiments, and a certain degree of autonomous decision-making capabilities to leap from human social contracts to real "Smart Contract".

Data: an Important Production Factor in a Smart Society. On the one hand, data as a resource, like land, labor, capital, and other factors of production, drives global economic growth and the social development worldwide. On the other hand, data as a constructive force of social relations is the core of the time. Together

---

[16]Smart society is a highly connected, highly digital, highly accurate computing, highly transparent and highly Smart Society (Wang Yukai. Smart Society Forces State Governance to Be Smart, *Information China*, 2018, No. 1, pp. 34-36).

with material and energy, it has become one of the three essential elements for human activities in the natural world, and has changed from a symbol describing things to one of the essential attributes of everything worldwide. However, in a centralized social system, massive data is usually in the hands of a small number of people, such as the government or large enterprises, which only "speak" for a small number of people. Its fairness, authority, and even security may not be guaranteed. But blockchain data, stored in highly redundant distributed nodes, is in the hands of "everyone", thus achieving "data democracy". Meanwhile, with its reliability, security, and immutability, blockchain can fully realize data sharing and data computing on the premise of ensuring data trust, data quality, and data privacy security, supporting the applications of smart society in data quality and sharing. First of all, the immutability and traceability of blockchain enables the collection, transaction, circulation, and every step of computational analysis to be retained on blockchain. In the blockchain network, anyone cannot tamper with and modify data or create false data at will, which endows data credibility and quality a certain degree of credit endorsement and helps artificial intelligence to carry out high-level modeling, thus allowing users to get a better user experience (Xizi 2018). Secondly, based on homomorphic encryption, zero-knowledge proof, differential privacy, and other technologies, blockchain can realize the protection of data privacy in multi-party data sharing, enabling data owners to carry out data collaborative computing without disclosing data details. Finally, the incentive mechanism and consensus mechanism based on blockchain help expand the channels of data acquisition. On the premise that blockchain cryptography technology guarantees privacy and security, it is possible to collect the required data from all participants of the blockchain network worldwide based on pre-agreed rules. As for the invalid data that don't conform to the pre-agreed rules, the consensus mechanism will be adopted to eliminate them, so as to ensure reliable and high-quality data to guarantee computing and modeling, improve the value of blockchain data and expand its utilization. In short, the blockchain can further standardize the use of data, refine authorization scope, help break data isolation, form block data aggregation and flow, and achieve secure and trust data sharing and opening under the premise of protecting data privacy.

Algorithm: the Core Engine of Smart Society to Enhance Social Productivity. Smart contract is executed on multiple nodes. It relies on algorithms to build trust instead of being subject to man's will, which avoids the danger of being controlled by subjective will as in traditional contracts and laws, so as to make the execution result fairer. "Mastering data means mastering capital and wealth; mastering algorithm means mastering the right to speak and regulate" (Changshan 2019). A Smart Contract in blockchain is essentially a piece of code that implements an algorithm. Since it is an algorithm, artificial intelligence can be embedded, thereby making blockchain Smart Contract smarter. As Lawrence Lessig said, "In the future, code is both the greatest hope and the greatest threat to freedom and liberal ideals. We can design, program, and build a cyberspace that preserves the core values that we believe in, or we can let those values evaporate in this cyberspace. We have neither a middle way nor a sure card. Code is not discovered; it is created by man" (Lessig 2009). In addition, Smart Contract is another area where it is difficult for blockchain systems to cut ties with the law. "Blockchain smart contract is not only a technology, but reflects the changes and adjustments of the parties' rights and interests. Belonging to the object of legal regulation, it has both technical and legal significance" (Shaofei 2019). In fact, the Smart Contract code reflects the elements and logical structure of blockchain smart legal contracts. Even if there is no direct text of blockchain smart legal contracts, blockchain smart legal contracts can be confirmed through underlying rules or protocols of blockchain Smart Contract technology, operation mode, process and result of Smart Contract (Shaofei 2019). However "we need to determine and define what kind of social contracts would require 'code law' more", (Swan 2016) and whether this algorithm contract can be a contract in legal sense needs further demonstration. Therefore, "on the basis of adhering to the theory of interpretation, the current law needs to draw useful experience from the legal adjustment of electronic contracts, and use traditional contract laws to determine the nature of smart contract issuance and code execution, so as to provide an institutional framework for the laws to respond to Smart Contract" (Jidong 2019). In addition, the iteration and operation of algorithm need the support of strong computational power. In the future era of "Artificial Intelligence & Internet of Everything", technologies, whether it is the Internet of Things, big data, or artificial intelligence, have extremely high requirements for data storage and computing, e.g., the integration of blockchain and cloud computing. On the one hand, the basic services of cloud computing help promote the R & D of blockchain. On the other hand, decentralization of blockchain, immutability of data and other technical features help tighten the control of cloud computing over credibility, security, and other aspects (Xiangyu 2018). In this development, computing power is productivity, data is the means of production, and this form of blockchain organization computing power constitutes a new social production relationship.

Real Value of Blockchain: Scenario Applications. The application of blockchain technology has extended to digital finance, Internet of Things, smart manufacturing, supply chain management, digital asset transaction, and other fields. Smart contract is no longer a mere technology component of the blockchain system, but an increasingly independent new technology under research and application. At present, neither Bitcoin nor Ethereum, nor any other similar digital currency, can replace legal tender. Because they don't have the characteristics of "universal equivalent", and cannot be natural currencies, nor can they accurately describe the value of commodities in the market or offset the harm caused by inflation and deflation. In essence, digital currency based on blockchain technology is a kind of "reward points" in virtual space, with certain investment value only in specific ranges.[17] Therefore, the blockchain technology should be applied to real-life scenarios as soon as possible, rather than staying in the crazy hype of digital currency. The practical scenario application of blockchain technology can boost the mutual progress of social economy on a more pragmatic basis. Features of blockchain such as "decentralization", "openness", and "information immutability" bring a brand-new social relationship that does

---

[17]Nan Yunlou. The Value of Blockchain Lies in the Application of Practical Scenes. *Shenzhen Special Zone Daily*, January 30, 2018, p. C3.

not require endorsement by a third party and is absolutely reliable. And "Blockchain Plus" will also bring about new application scenarios and changes in production relations. The application field of blockchain is based on its two basic attributes, namely, financial and regulatory attributes. Its financial attribute means the blockchain is a "trust machine", which establishes trust mechanism through algorithms. All financial services could partially or fully adopt blockchain technology in principle. Of course, it is necessary in implementation to consider the costs and promotion resistance involved in inventory and increments. For example, in credit investigation, supply chain finance, lending, bills, securities, insurance, and other enterprises, blockchain has broad application prospects. In government affairs, blockchain can be applied in regulation, approval, data sharing, arbitration, licenses, and so on. In industry, blockchain can be applied to such fields as supply chain certificate, electronic contract, quality management, logistics, and product traceability. In terms of people's livelihood, blockchain can be applied to the fields of integration of culture and tourism, digital identity of new retail industry and product traceability.

4.2.3 Digital Citizens and Social Governance

Blockchain is changing our world in a more rapid and fierce way, pushing people to migrate to the digital world. The arrival of the digital world has given birth to "digital citizens". As a major component of digital nation, "digital citizen" is of great significance to innovative social governance and public services. As a governance technology, blockchain differs from traditional political agendas in that its governance rules are embedded in algorithms and technical structures. Therefore, it is urgent for us to break through the traditional thinking habit of single physical space and establish the new idea and new thinking of dual space and smart governance. We need to design regulation schemes according to the production, life and behavior logic of dual space, while integrating and adding modeling, algorithm, code, and other ways, so as to tackle the problems and challenges of grassroots governance of smart society, and build an effective order of smart governance (Changshan 2019).The governance architecture based on blockchain technology "not only copes with mutual trust among the public, but enables the public to play an active role in social governance, form a collaborative social governance model, maximize public interests, and promote good social governance" (Yanchuan 2019).

Digital Citizens. "Digital citizens" refer to digitalized citizens or citizens' digitization. It is the reflection of citizens in the digital world, the copy of citizens in the physical world, the digital presentation of citizens' responsibility, rights and interests, and an important part of individual citizens (Jing 2019). In the cyber-world, it is difficult to implement the real-name system, which brings great difficulties to the governance of cyberspace and integrity system. The core reason is the lack of citizens' digital identity, just like missing identity cards in the physical world, which makes citizens unable to prove "I am myself", let alone to enjoy rights and fulfill obligations. By building a unified citizenship authentication platform and identity authentication system, "isolated islands of information" can be connected so as to endow citizens unimpeded digital identities in the digital world. Only a credible digital identity issued and managed by government departments, with related laws and regulations that are amended and improved, can provide citizens with a unified entrance to participate in social governance based on real-name authentication environment. In addition to providing easy access to identity (identification), blockchain, with its distributed ledger data and immutable database, allows the one who demands the proof of identity information and the one who provides the proof of identity to take what he needs once the identity of the person is on the chain, so as to build a trusted digital citizenship. Relying on trusted digital identity, in the new space where the physical world and the digital world merge, a set of comprehensive systems is established around the identification of various subjects, which can realize the secure and credible authentication of identity in real and internet scenarios (Jing 2019). "Digital citizens" give birth to "diverse" subjects of social governance and construct a bottom-up public service innovation system driven by technologies. They make citizens willing and active to participate in social governance and form synergy with the top-down governance system of the government, so as to transform the management mode of single subject of the government to the collaborative governance mode of multiple subjects. "Digital citizens" make people's identity in the digital world identifiable and verifiable, people's behavior traceable, and the supervision of people more convenient and systematic. "Digital citizens" living in the digital world are bound to regulate their own behaviors more autonomously.

Blockchain Innovates Livelihood Services. General Secretary Xi Jinping emphasized that it is necessary to explore the application of "Blockchain Plus" in people's livelihood, and promote the use of blockchain technology in education, employment, pension, targeted poverty alleviation, medical health, commodity security, food safety, public welfare, social assistance, and other fields to provide people with smarter, more convenient, and higher quality public services. The application of "Internet Plus" has brought great convenience to people, so it is not hard to imagine "Blockchain + people's livelihood" also has broad prospects in application. Blockchain technology can be applied to the recording, notarization, and service of important data of people's livelihood to ensure the authenticity, reliability, and immutability of data, so as to rebuild social credibility. For example, blockchain can be applied to the fairness and transparency of poverty alleviation, employment, and social security; recording and preserving private data of personal health; tracing food supply; recording student scores and academic certificates.[18] The application scenarios of "Blockchain Plus" in people's livelihood are far more than these. Theoretically, all the people's livelihood services that require trust, value, and cooperation can be provided with complete solutions by blockchain technology with innovative practices of "rich imagination", including certificate application, business processing, medical expense reimbursement, provident fund distribution, microfinance investigation, judicial trial evidence chain, and notarization. For example, in the food and

---

[18]The Information Office of Guiyang Municipal People's Government. *Guiyang Blockchain Development and Application.* Guizhou People's Publishing House, 2016.

medicine safety that people care about most, blockchain technology can help build a database that is traceable and immutable in the whole process, so as to accurately monitor food and medicine safety. As another example, in medical insurance, one of the biggest pain points used to be the inability to fully and accurately grasp the real health status of the insured, because the insured may go to different hospitals for examination and treatment, or get insured in different insurance companies. With the help of blockchain technology, data between insurance companies and hospitals can be shared on the basis of data privacy, security, and reliability, thus bettering medical insurance industry.

Blockchain Empowers Social Governance. "In social governance and public services, blockchain has a wide range of application with the potential to effectively improve the digitalization, intelligence, refinement and legalization of social governance." (Fuwen 2019) "Generally speaking, traditional social governance belongs to unitary governance model, of which the cost of public participation in social governance is relatively high" (Yanchuan 2019). The consensus mechanism and Smart Contract in blockchain create transparent, reliable, efficient, and low-cost application scenarios. It can build smart mechanisms of real-time interconnectivity, data sharing, and linkage collaboration, so as to optimize government services, city management, and emergency support, and improve governance efficiency. As the underlying support, blockchain technology can change public participation in social governance, and promote social coordination and democratic consultation. The distributed and peer-to-peer technology of blockchain can realize multi-agent participation in social governance and help each social subject to construct trust mechanism under government guidance. Blockchain is conducive to adjusting the relationship between people, thus changing the previous social governance model dominated by the government, so people can participate in social governance and deeper in economic and social development and reforms (Dong and Chenhui 2019). Sovereignty blockchain endorses the principle of consensus to social contract theories. The consultative democracy advocates the concepts of tolerance, equality, rationality, and consensus. In practice, to transform these concepts into daily practice, we must rely on modern science and technology. The distributed form, trustworthiness, and other technical features of Sovereignty blockchain are conducive to better explaining the existing political system, and may even activate some of its "deeply buried" functions, thus leading to a more stable path. As a general ledger, Sovereignty blockchain, not simply a currency transaction accounting system, serves as a platform for people to reach agreements and consensus on everything without third-party intermediaries. Sovereignty blockchain technology is conducive to eliminating the inherent defects in the trust basic model and giving full play to the role of consultative democracy. Under the context of sovereignty blockchain, the elements of consultative democracy implied in the political consultation system, such as "rational consultation", "equal respect", and "consensual orientation" will be activated and strengthened, and may even lead traditional consultation controlled by elites to some extent to a democratic practice that is more inclusive of public participation (Key Laboratory of Big Data Strategy 2017). Distributed consensus is the foundation of Smart Contract autonomy where every node participates in governance, thereby achieving "autonomy" (Ke 2019). It can be said that the Smart Contract based on the underlying technology of blockchain will reconstruct the co-governance pattern of the state, government, market, and citizens, and promote the genuine entry of human society into a contractual society in the multi-governance network social governance system.

4.3 Traceable Government

In *The Social Contract*, Rousseau stated that an ideal society is based on the contractual relationship between people rather than between people and government,[19] that is, the government is the executor of the sovereignty, not the sovereignty itself. However, the "direct democracy" advocated by Rousseau is also out of frustration that there is no basic trust between people, because once human will is represented, it will eventually be distorted. In the digital society, with the intervention of blockchain Smart Contract, technology creates machine trust, which changes the contractual relationship between people from original government trust endorsement to technology trust endorsement. Machine trust eliminates the intervention of human factors, and makes information data in the entire process immutable, irrevocable, verifiable, traceable, and accountable, so as to tackle the trust "blocking" of digital government affairs in the cyberspace. It builds a true and undeniable trust foundation for "digital certificate", "digital contract", and "digital system". It can be said that the governance paradigm based on blockchain and Smart Contract will be a new manifestation of social contract in the era of digital civilization. It can effectively transform government roles and functions, improve the transparency of government organizational structure, governance, and service processes, achieve social fairness and justice, accelerate transformation to digitization, build a credible digital government, and steadily empower the modernization of state governance system and governance capability.

---

[19]Rousseau, in his *The Social Contract*, discusses the principle of government organization system: "Government is an intermediate body established between the subject and the sovereign in order that the two may be suited to each other, and that it is responsible for the execution of the law and the maintenance of social and political freedom." In his view, governments are composed of humans who are the consent of the people as sovereigns, who are the public servants of the people; it is natural for the people to restrict, correct and replace the government; The people have an eternal revolutionary right to government; The power of the state is derived from the transfer of individual rights, and the power of the government is derived from the delegation of the sovereign, so the power of the government must always serve the people. Rousseau's radical bourgeois democracy theory is the theoretical basis of the French Revolution, especially the Jacobin dictatorship.

4.3.1 The Regulation of Government Power

Rousseau believed it is not a contractual act that constitutes the government system, but a legal act; the holder of administrative power is not the master of the people, but the employee of the people. On the surface, the ruler is only exercising his power, but this power is easy to spread (Rousseau 2012). In the entity of human society, the government or leaders shoulder the rights and obligations of formulating and implementing the social contract, which is a centralized system. The executor of a digital society, however, is no longer a centralized "government". Distributed group decision-making enables each individual to enjoy the crystallization of group experience, so as to continuously carry out the benign iteration of individuals and groups. Meanwhile, the history of blockchain can be traced, so the records of everyone involved in decision-making can be preserved, immutable and accepted by everyone's supervision and management. The contract of the entire digital society will be more transparent, democratic, and open. In particular, big data changes the track of traditional public power operation, recreates the subject and object of power, and opens a window for us to re-understand power. "Data is power, power is data". Power is endowed with the attribute of data. Data decentralization and checks and balances have become the new normal. The features of decentralization, openness, and sharing become the main characteristics of power (Key Laboratory of Big Data Strategy 2016). When traditional power structure system is broken by data dimension, power boundary is adjusted by data dimension, and power logic is rewritten by data dimension, power can be locked into the cage of system. These changes will eventually be transformed into a kind of visibility and positive energy of state governance, laying a solid foundation for real "data governance".

Blockchain Truly Realizes Power Datamation. Power datamation is the premise of power segmentation, measurement, calculation, reorganization, and standardization. It is the guarantee to make the operation process of every power standardized, transparent, quantifiable, analyzable, and controllable (Key Laboratory of Big Data Strategy 2017). However, there are risks such as deletion and tampering of data, which makes it difficult to guarantee the dynamic tracking supervision and intelligent analysis of the whole process of power. Blockchain can technically guarantee data retention, data collection, data correlation analysis, and data intelligence, and realize power datamation in real sense, so the data of power can be public, sources found, destination traced, and responsibility investigated, which provides a new scheme for the supervision of government power. Firstly, data leaves traces behind. Blockchain technology enables power data to leave traces behind everywhere, and ensures the trace of data is always recordable, dynamically searchable, and traceable throughout the whole process. The multi-node co-participation and co-supervision mechanism on blockchain that determines data will not be "hidden" (records not uploaded). Secondly, data aggregation can be achieved. Data aggregation is the premise of data correlation analysis and the basis of realizing data value. Blockchain ensures all nodes in the network participate in and jointly verify the authenticity of data in the same network. The data generated based on the consensus of the whole network is reliable and immutable structured data. Thirdly, data correlation analysis becomes possible. Blockchain is a rope network structure, which associates data between chains with the formation of a network. The Consensus Mechanism of blockchain and the characteristics of transparency and de-trust determine that the data recorded on blockchain has a good basis for data analysis. Finally, data intelligence is enabled. Blockchain Smart Contract provides support for the intelligence of power data. Once the power data is recorded on blockchain, intelligent risk warning based on intelligent matching of power data can be realized.

Blockchain Tackles Governance Hard Nuts in Government Power Regulation. Due to the low transparency of government information, low participation of citizens in politics, (Jianping and Yun 2017) weak public services of government, monopoly, and abuse of government power, the lack of government trust has become increasingly serious. Government trust based on blockchain should be built layer by layer from the front to the end of government functions. Blockchain should be applied to government public services, multi-party participation, information sharing, supervision feedback, etc. The initiative of government trust will rely on technical tool; complex government trust challenges will be simplified; the ability of the government to perform functions and efficiency will be enhanced; multicenter benign interaction will be promoted; government affairs openness, information sharing, all-round supervision, and feedback will be realized; a blockchain-based government trust network with clear logic will be built. Then a government trust ecosystem based on blockchain will be formed (Feifei and Xuedong 2019). In addition, in the blockchain-based public service network, the status of all stakeholders tends to be equal, with the government department only one member of the network. This will lessen the government control over public power in public services, reduce public reliance on government departments, and weaken the management authority of government departments to some extent. In the future, the deep integration of big data, artificial intelligence, blockchain, and other information technologies will be conducive to cross-regional, industry-wide, cross-administrative classes, full life cycle management blockchain regulatory service platform, the transformation of government functions and regulatory innovation, optimizing the government business process, including regulatory agencies and government agencies, so as to make the Open Government to be positive, transparent, and credible.

"Data Cage" is a Pioneer of Technological Anti-corruption. In 2015, "Data Cage" was introduced to the public as "an innovative move to address the major issue of supervision restricting power". "Data Cage" took the lead in carrying out pilot applications in government departments such as the Municipal Housing and Urban-Rural Development Bureau, the Municipal Traffic Management Bureau, which have relatively centralized administrative powers and are closely related to the lives of people. The exploratory and innovative model of anti-corruption built on the "data cage" project was launched. "Data cage" is essentially to put power in the cage of data and let the power be exercised in the sunshine. It is a self-organizing system engineering with informatization, datamation, self-processing, and amalgamation of power exercising and power restriction as the core (Fig. 4.4). It uses big data to realize the transparent management of government negative lists, power lists, and

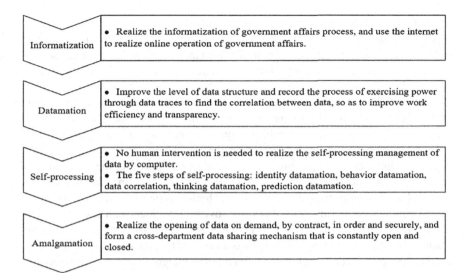

Fig. 4.4 The "Tetralogy" of public power governance

responsibility lists, optimize and reengineer administrative management processes, enhance government management and public governance methods, simplify government decentralization and administration according to law, and respond to the "inaction, slow action, and disorderly action" on the part of leading cadres and public officials (Key Laboratory of Big Data Strategy 2017). To lock up power through the cage of institutions means digitizing all the process of power operation, documenting the track of power operation, supervising, regulating, checking, and balancing power in the cage, so as to ensure the proper exercise of power.[20] Specifically, the "data cage" works for such functions as administrative examination and approval services, unified management of registered items, data sharing and collaborative approval by building a unified platform related to administrative examination and approval services. It improves public trust in government authorities. Meanwhile, it uses the cloud platform to effectively monitor market behaviors, so that the dishonest behaviors in the market can be punished and market behaviors can be standardized.[21]

Blockchain Becomes an Essential Part of Regulatory Technology. "Data cage" is conducive to the application of big data projects for operation supervision, performance evaluation, and risk prevention of government power, but it also faces some problems. The first one is the passive application of "data cage". The application of "data cage" is used to motivate civil servants at all levels to have conscious actions and set up new rules and orders. Second, the application of "data cage" is relatively

---

[20] Jia Luochuan. Reflections on Building an Anti-Corruption System Cage for Prison Police, *Crime Prevention and Control and the Construction of a Safe China—Proceedings of the Annual Meeting of the Chinese Society of Criminology*. China People's Procuratorate Press, 2013, p. 91.

[21] Key Laboratory of Big Data Strategy. "Data Cage": A New Probe into Technical Anti-corruption Key. *China Terminology*, 2018, No. 4, p. 77.

independent, and there is still a risk that business data will be tampered with. Based on this, blockchain uses codes to build a trust method at the lowest cost, namely, machine trust. There is no need to try your best to see through "banana oil", to get endorsement by the government, and to worry about the unfairness and corruption of the system, which offers solutions to the application of "data cage". Firstly, the government should establish a "data cage" regulatory platform based on sovereignty blockchain and a "data cage" blockchain covering all the nodes of government departments. It facilitates the formation of immutable encrypted records of important powers' operation data of departments on the blockchain, promotes mutual supervision of power operation, and establishes a comprehensive evaluation system for the application of "data cage" of departments, so as to make the "data cage" more firm, transparent, and binding. Secondly, the government should build a system of civil servants' disciplined compliance and integrity based on sovereignty blockchain and record key information such as civil servants' compliance and performance on the blockchain. Each node of the department jointly verifies and audits to establish an immutable civil servant "integrity chain". Each department inquires integrity records of civil servants according to its authority. On the basis of such a system of civil servants' discipline and integrity, a value incentive mechanism for the application of "data cage", such as workload incentive and "thumb up" mechanism, will be established taking the system as an important evidence for assessment, appointment, reward, and punishment. Inspection integral is formed based on the application of "data cage" of cadres and discipline inspection committee members, and is recorded in the system of civil servants' discipline and integrity.[22] The combination of blockchain technology with "data cage" standardizes and restricts the effective operation of power, and ensures the regulation of power leaves traces everywhere.

4.3.2   Traceability of Government Responsibility

People have a love-hate relationship with governments. Originally, the governments represent and safeguard people's interests, and government officials are public servants. Without the governments, nations will face a series of risks such as collapse of political order, failure of market mechanisms, and decline of public services. As the government grows with its own scale, especially when it makes peace with power, certain evils will be revealed, including official corruption and government capture, such as the excess of power and dictatorship. Therefore, In *Two Treatises of Government*, British scholar John Locke regarded the government as a "necessary evil" and clearly proposed the concept of "limited government", that is, "constitutional government" means "limited government". Marx and Engels called the nation the "drag" and "tumor" of a society. "At most it's nothing more than a scourge inherited by the proletariat after the victory of struggle for class rule". To put it bluntly, the

---

[22]The Information Office of Guiyang Municipal People's Government. *Guiyang Blockchain Development and Application.* Guizhou People's Publishing House, 2016.

government's irresponsibility, dishonesty, and immorality damage social credit to a considerable extent. Its essence is power misconduct. Responsibility is the main quality that contemporary government should possess, and it's also a very noticeable characteristic of contemporary government. Only a government that is responsible to its citizens and has limited power is a responsible government. "A responsible government means the government can actively respond to, meet and realize the legitimate demands of citizens. A responsible government is required to assume moral, political, administrative and legal responsibilities. Meanwhile, it also means a set of control mechanisms for the government" (Chengfu 2000). A traceable government based on blockchain can technically build a responsible government where all the sources and directions can be traced, and all the responsibilities can be investigated. It addresses problems, e.g., opacity, low efficiency, and difficult supervision. It can fundamentally reconstruct the benign interaction between governments, enterprises, and citizens, and restore the trust of governments.

Government Responsibility and Responsible Government. A modern representative democratic government is essentially a responsible government. "In modern political practice, the responsible government is not an expression of will, but a political principle and a government responsibility system based on this principle" (Bangzuo and Yucheng 2003). According to the principle of people's sovereignty, the origin of state power lies in its people. However, it is impossible for the people to manage national and social public affairs. It is necessary to produce subjects of state power that can represent the people's interests to manage national and social public affairs through certain rules and procedures and according to the will of the people. The government is a very important part of this subject of power. According to principal-agent theory, the power of governments comes from the people. When the government acquires power, it also assumes corresponding responsibilities, so the government is only legitimate when it truly assumes the responsibilities directly or indirectly given by voters. Responsible governments mean the governments should be responsible to the legislature, to the laws enacted by the legislature, to respond, to meet and realize the legitimate demands of citizens, and to use its power in a responsible way. Behind every power exercised by the government, there is a responsibility. There is no doubt the responsible government is responsible for all the people and their representative organs based on whom the government has derived according to certain rules and procedures. According to the principles of democratic politics and legal administration, first of all, the government should assume the responsibilities stipulated by the Constitution and laws that reflect the will of the people, and assume the responsibility of exercising power in violation of the Constitution and laws. Second, the principles of democratic politics and representative system require administrative officials to formulate public policies in accordance with the will of the people and to work for the implementation. If, in policy formulation and implementation, administrative officials make wrong decisions, violate public opinions, fail to perform their duties, or commit dereliction of duty, they shall bear corresponding political responsibilities. Thirdly, national civil servants working in government agencies should be the models who take the lead in observing disciplines and laws. They should fulfill their duties, be diligent to serve the people, and be honest, fair, and upright. In addition, they shouldn't use power for personal gains. These are duties and responsibilities of civil servants. If civil servants fail to perform their duties, they should not only bear the responsibility for breaking the law and discipline, but be condemned by social morality and bear moral responsibility for violating administrative ethics (Fangbo 2004).

Traceability Mechanism of Blockchain. The traceability mechanism refers to the mechanism that can track information in an all-round way. It takes the point-to-point supervision network as the basic carrier to timely trace the data changes, transactions, and other information of each node. Once there are quality and safety concerns, the main responsible person can be traced back in time and the responsible person shall bear relevant responsibilities (Key Laboratory of Big Data Strategy 2017). The logic of the traceability mechanism is based on three elements: information, risk, and trust. The corresponding information risk responsibility mechanism is established on the three elements. Regulators clarify the key role of information, risk, and trust in supervision and management, and meanwhile construct an effective mechanism of information, risk, and trust. This mechanism is an effective way to change the previous slapdash supervision model. It uses these three elements to integrate supervision and management into a standardized, orderly, basically controllable and predictable performance framework. The block header of each block on the blockchain contains the hash value of the transaction information of the previous block, so the starting block (the first block) is connected to the current block to form a long chain, and each block must follow the previous block in a chronological order. The structure of "Block + Chain" provides a complete history of a database. From the first block to the newly generated block, all historical data is stored on the blockchain. This enables us to search for each piece of data in the database, and every piece of transaction data on the blockchain can be traced back to the source through the structure of the blockchain. In the blockchain, data will be monitored by the entire network in real time, and any attempt to tamper with or delete information will be detected, recorded, and rejected by the blockchain. The whole traceable data can be clearly defined after ensuring the timeliness, accuracy, and effectiveness of data in each link (block), thus truly achieving "production recorded, process regulated and responsibility traceable".

A traceable Government Based on Blockchain. Blockchain is suitable for multi-state and multi-link, which requires the joint participation of all parties for collaborative completion, and where the parties don't trust each other and can't use a trusted third party. The establishment and maintenance of government trust require information openness, transparency, credibility, immutability, traceability, power restriction, and multi-party collaborative participation. The applicable criteria of blockchain are exactly in perfect accord with the government's value pursuit of "publicity + trust" (Feifei and Xuedong 2019). The government responsibility mechanism of multi-layer cooperation and multi-head interconnection based on blockchain technology enables each node on the blockchain to have the right and obligation to jointly monitor and maintain the data on the chain, collect traceable and immutable data, and implement the principle of "whoever produces and changes the data shall be the responsible person". This avoids the risks caused by massive data, thus improving the transparency and democracy of government decision-making and helping construct credible and traceable governments. In addition, the immutability and traceability of blockchain make all information activities on the chain searchable and traceable, form credit files of all members automatically, and maximize the supervision of social members over the government and other contact objects (Yi and Yi 2019). Therefore, with the establishment of a multi-agent supervision and feedback link through blockchain, each subject should not only be responsible for data input, but jointly share the supervision responsibility of blockchain. With decentralized, immutable, and traceable blockchain technology, a public decision-making responsibility mechanism of "self-organized operation of government service platform with government assistance" is to be constructed.[23] It can be said that under the contractual constraint of public supervision, blockchain provides the technical foundation for public participation and distributed autonomy to guarantee trust, constantly forces the government to form a perfect traceable responsibility mechanism, and lays a solid foundation for the future digital government.

Blockchain Empowers Traceability of Public Welfare. In COVID-19 epidemic, the Red Cross Society of China Hubei Branch and the Red Cross Society of China Wuhan Branch were questioned because the information release of materials and donations was not timely, open, and transparent, which are also the pain points and difficulties faced by current public welfare charities. Distributed, difficult to tamper with, and traceable, blockchain technology can effectively streamline the complex process and prevent black box operation in traditional public welfare and charity. Specifically, an open, transparent, and traceable charitable donation platform can be established through blockchain. Blockchain distributed ledger and consensus mechanism can be used to record information related to the donation and adoption, including information about the donor, recipient, intermediary institutions, donated materials, and other entities, as well as information about donating, distributing, receiving, and confirming materials to recipients. A unique blockchain identity is issued to each participant, real-name authentication carried out, each link signed by the participant. These actions help prevent forgery, fraudulent claims, and others. Meanwhile, the identity of institutions on the chain is transparent with all transactions broadcast online, and each node recorded on the ledger, so blockchain is traceable. People who are concerned can inquire and trace every transaction, trace it to the responsible person point to point, verify the authenticity of data records, trace data records back to the source, make the subject responsibility clear, ensure the openness and transparency of public welfare projects, thus reshaping the credible public charity system.

---

[23] Jiang Yuhao, Jia Kai. Research on Public Decision-making Responsibility Mechanism Reform Based on Big Data under the Blockchain Technology Path. *E-government*, 2018, No. 2, p. 32.

4.3.3 Paradigm of Digital Government

It was clearly stated in the Fourth Plenary Session of 19th Central Committee of the CPC: "Establishing and improving the systems and rules of administrative management by using Internet, big data, artificial intelligence and other technical means; promoting digital government, strengthening the orderly sharing of data, and protecting personal information in accordance with the law." The digital government has been a major approach and key choice to "innovate administrative mode, improve administrative efficiency, and build a service-oriented government that people are satisfied with".[24] Governance of digital governments is conducive to state governance, social governance, and government governance capacities. Up to now, digital government governance is still a relatively vague and indefinable concept of public management theory. According to domestic scholars who have theoretical interpretations and academic researches on digital government governance from different perspectives and fields, we can understand the logic of digital government governance from the theoretical perspective and goals. From the theoretical perspective, human society has evolved from an agricultural society to an industrial society, and then to a digital society. The increasingly high degree of informatization has promoted the government to change the governance method, that is, from the traditional representative interaction and one-way control to digital negotiation and mutual consultation, construction, and sharing. In terms of the goal, as a new way of state governance, digital government governance focuses on the linkage change and shared development between digital government and other governance bodies. Its goal comes to shift from leveraging government governance to creating common value (Ling 2019). The key to the digital government is not "digitization" but "governance". The blockchain-based governance architecture can provide a basic architecture of data trust for digital government governance.

Sovereignty Blockchain Becomes the Digital Infrastructure for Government Governance. "In the digital government system, sovereignty blockchain serves as the digital infrastructure of government governance. It will combine technical rules with laws and regulations to complete sci-tech supervision, as well as law enforcement and governance. On this basis, digital government will perform economic, political and social functions under the new definition, and complete the construction and implementation of social governance system".[25] With the advent of the era of Internet of Value, the innovation of theory and technology is giving birth to a new digital government system and governance paradigm. If Information Internet has tackled information asymmetry and narrow expression, and expanded the democratic scene, then blockchain provides a set of democratic operation mechanism on this basis,

---

[24]Chen Jiayou, Wu Dahua. Building a Digital Government and Improving the Modernization of Governance Capacity. *Guangming Daily,* December 9, 2019, p. 6.

[25]Capital Wings. When Talking about the Digital Government under Blockchain, What Should We Talk about. Odaily. 2018. https://www.odaily.com/post/5133065.

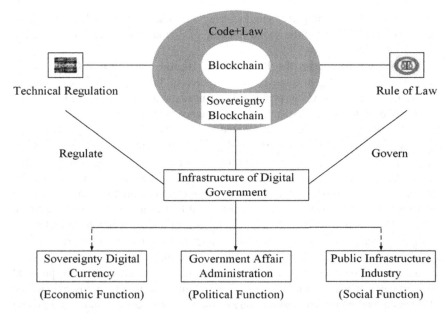

Fig. 4.5 Government governance system based on sovereignty blockchain. *Source* Capital Wings, When Talking about the Digital Government under Blockchain, What Should We Talk about, *Odaily*, 2018, https://www.odaily.com/post/5133065

which enables the hierarchical internal structure of government and the government-centered social governance system to be adapted, and promotes blockchain "consensus" into Sovereignty blockchain "co-governance" (Fig. 4.5). In the early stage of Sovereignty blockchain, the application of digital government focuses on upgrading administrative management or public service information system by using blockchain technology. It mainly covers digital identity registration, credit authentication, data service, etc., in the form of alliance chain led by relevant national institutions. As the Sovereignty blockchain matures, the economic functions of digital governments are expected to be first realized due to the natural combination of blockchain, credit and value. Sovereignty digital currency is an essential manifestation of the economic functions of digital governments. Sovereignty digital currency not only has the advantages of digital currency in openness, transparency, and traceability, but those of the sovereignty currency in security, stability, and authority. It is expected to become one of the mainstream forms of currency in the future. However, confined by the efficiency and security of the current blockchain technology, the huge impact and high risk of issuing sovereignty digital currency, the governments of all nations currently have a positive and cautious attitude toward its research. Government administration involves system management within the government, which demands high the security and stability of sovereignty blockchain. Therefore, the political functions of digital government will be realized next to economic functions. Public infrastructure is mainly centered in industry, manufacturing and other fields, but less to be integrated with blockchain, so the social function of digital government is expected to be realized at last.

Digital Twin Innovates the Paradigm of Digital Government Governance. "After the era of intelligent Internet, the formation of dual space of physics (reality) and electronics (virtuality) has profoundly changed the previous mode of production and lifestyle" (Changshan 2019). Urban governance system is also undergoing major changes under the trend of service-oriented digital government. It is urgent to establish new concepts and new thinking of digital government governance, and shape a new orderly environment for government governance according to the production mode and lifestyle of dual space, and the logic of a smart society. "Driven by the new generation of information and communication technology and the reform of urban governance system and mechanism, current grid-based fine management model will gradually evolve into a highly intelligent and autonomous model based on digital twin".[26]Digital Twin is a new manufacturing mode driven by data and model and supported by digital twin objects and digital thread. Through real-time connection, mapping, analysis, and feedback of assets and behaviors in the physical world, it can achieve maximum closed-loop optimization of total industrial factors, whole industrial chain, and entire value chain.[27] "Both public domain and private domain have broken through the traditional meaning and scope of physical space and are continuously expanding and extending to the virtual space. Moreover, people's behaviors and social relations have also undergone profound changes in the virtual and real isomorphism" (Changshan 2020). People, objects, and events in the physical world are completely mapped to the virtual world, and can be fully monitored through intelligent processing. Intelligent processing helps master the physical world and establishes connections between the virtual world and the physical world by adjusting digitized elements, thus influencing the physical world. The real world and the virtual world coexist and blend with each other. Digital Twin city is a complex and comprehensive technological system supporting new smart cities, and an advanced model of steady innovation in urban smart operation. It is also a future form of urban development in which the physical city in the physical dimension and the digital city in the information dimension coexist and integrate with each other.[28] In the future, the whole world will generate a digital twin virtual world based on the physical world, where human and human, human and objects, objects and objects will deliver information and intelligence through the digital world (Fig. 4.6).

Blockchain Builds a New Order of Digital Civilization. Key technologies such as blockchain, artificial intelligence, quantum information, 5G, and Internet of Things build a digital twin world, promote the digital government, and breed a new order of

---

[26]Zhang Yuxiong. On the Reform of Digital Twin City Governance Model. Sohu. 2018. http://www.sohu.com/a/224351264_735021.

[27]Liu Yang. Digital Twin Key Technology Trends and Application Prospects. The China Academy of Information and Communications Technology. 2019.

   http://www.caict.ac.cn/kxyj/caictgd/tnull_271054.htm.

[28]Xu Hao's speech at the Second Smart China Expo "Intelligent Application and High-Quality Life Summit": "Digital Twin, Paradigm Change of Smart City". August 26, 2019.

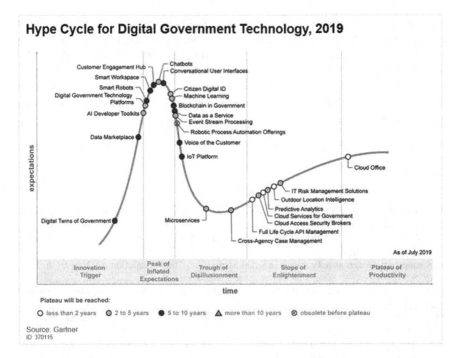

Fig. 4.6 Hype cycle for digital government technology (up to July 2019). *Source* Gartner, Hype Cycle for Emerging Technologies (2019), 199IT, 2019, http://www.199it.com/archives/958316.html

digital civilization. Kevin Kelly believed the world in the future is a world with virtuality and reality, and we hope to map the whole real world into a digital society. In the mirror world of Digital Twin, we interact with tools. To ensure the credibility and authenticity of uploaded data, we hope the data to be decentralized, so blockchain technology is needed. The technical advantage of blockchain can ensure the data invariability of Digital Twin program. Based on the programmability of blockchain Smart Contract, various virtual experiments, scene deductions, and result evaluations obtain the optimal decision of the system, which makes the digital city model fully emerge on the surface. As a twin, it runs parallel with the physical city. The virtuality and reality integrate with each other, so they contain infinite space for innovation. Therefore, the Smart Contract integrated the physical world with the virtual world, takes program code as the executor of the contract, makes the breach and dishonesty absolutely impossible, and provides a series of trust services for the digital government governance. The organic combination of blockchain and Smart Contract helps realize the intelligence of social relations and tackle the relationship between people and intelligent agents, thus generating value, which is an important direction in the future. The social relation built by Smart Contract will be a new relation of smart society, which will lead humans to the real contract society. In the future, key technologies represented by blockchain, artificial intelligence, quantum information, 5G, and Internet of Things will profoundly impact urban civilization construction and contribute to the birth of Digital Twin world. There is no doubt blockchain, as the underlying architecture, also provides ideas, mechanisms, and technology implementation solutions for urban civilization, which will accelerate human society to the new era of digital civilization. Moreover, digital civilization sets the development principles and direction for blockchain, and blockchain government will be given a new mission.

References

Wang Bangzuo, Sang Yucheng. On Responsible Government. *Digest of Party and Government Cadres*, 2003, No. 6.

Fang Biao. Smart Contract Boosts the Construction of Smart Society. *Chinese Social Science*, August 28, 2019, p. 7.

Ma Changshan. Governance Problems in Smart Society and Their Resolution. *Seeking Truth*, 2019, No. 5.

Ma Changshan. Recognition and Protection of "Digital Human Rights", *Beijing Daily*, 6 January 2020, p. 14.

Zhang Chengfu. On Responsible Government. *Journal of Renmin University of China*, 2000, No. 2, p. 75.

China Academy of Information and Communications Technology. Blockchain White Paper *(2019)*. China Academy of Information and Communications Technology website. 2019. http://www.caict.ac.cn/kxyj/qwfb/bps/201911/P020191108365460712077.pdf.

Sun Chongming. Turn Danger into Opportunity and Improve the Scientific Research and Innovation Ability of Enterprises. *Business China*, 2019, No. 4.

Huawei Cloud. Huawei Blockchain White Paper. echinagov. 2018. http://www.echinagov.com/cooperativezone/210899.html.

Yang Dong, Yu Chenhui. Application of Blockchain Technology in Government Governance, Social Governance and CPC Building. rmltwz. 2019. http://www.rmlt.com.cn/2019/1230/565266.shtml.

Cai Fangbo. On the Construction of Government Responsibility System. *Chinese Public Administration*, 2004, No. 4.

Chen Feifei, Wang Xuedong. Research on Building Government Trust Based on Blockchain. *E-government*, 2019, No. 12.

Gong Fuwen. Blockchain Empowers Social Governance. *People's Daily*, November 21, 2019, p.5

Guiyang People's Government Information Office. *Guiyang Blockchain Development and Application*. Guizhou People's Publishing House, 2016.

Mei Hong. Consolidating the Foundation of a Smart Society. *People's Daily*, December 12, 2018, p.7.

Cao Hongli, Huang Zhongyi. Blockchain: The Infrastructure of Building Digital Economy. *Cyberspace Security*, 2019, No. 5, p.76.

Wang Jian. *Being Online*. China CITIC Press, 2016, p. 34.

Liu Jianping, Zhou Yun. The Concept, Impact Factor, Changing Mechanism and Role of Government Trust. *Social Sciences in Guangdong*, 2017, No. 6, pp. 83-89.

He Jianqing. Financial Technology: Development, Influence and Supervision. *Journal of Financial Development Research,* 2017, No. 6.

Chen Jidong. The Legal Structure of Smart Contract. *Oriental Law*, 2019.

Wang Jing. "Digital Citizen" and Social Governance Innovation. *Study Times*, August 30, 2019, p.3.

Zhu Jiwei. Blockchain: The Cornerstone of Digital Finance. *Informatization Construction*, 2019, No. 7.

Xu Ke. Smart Contracts in Decisional Cross, *Oriental Law*, 2019, No. 3

Key Laboratory of Big Data Strategy. *Block Data 2.0: Paradigm Revolution in the Era of Big Data*. CITIC Press, 2016.

Key Laboratory of Big Data Strategy. *Block Data 3.0: Orderly Internet and Sovereignty Blockchain*. CITIC Press, 2017.

Key Laboratory of Big Data Strategy. *Redefining Big Data: Ten Drivers for Changing the Future*. China Machinery Industry Press, 2017.

Key Laboratory of Big Data Strategy. "Data Cage": A New Probe into Technical Anti-corruption Key. *China Terminology*, 2018, No. 4.

[US] Lawrence Lessig. *Code Version 2.0: And Other Laws of Cyberspace*, translated by Li Xu, Shen Weiwei. Tsinghua University Press, 2009.

Zhu Ling. The Realistic Dilemma and Breakthrough Path of China's Digital Government Governance. *People's Tribune*, 2019, No. 32.

Zhang Ling, Guo Lixin, Huang Wu, eds. *Crime Prevention and Control and the Construction of a Safe China—Proceedings of the Annual Meeting of the Chinese Society of Criminology*. China Procuratorial Press, 2013.

Liu Quan. *Blockchain and Artificial Intelligence: Building an Intelligent Digital Economy World*. Posts and Telecom Press, 2019.

[US] Reinsel et al. IDC: China Will Have the World's Largest Data Circle by 2025. asmag. 2019. http://security.asmag.com.cn/news/201902/97598.html.

[France] Jean-Jacques Rousseau. *The Social Contract (Bilingual Edition)*, translated by Dai Guangnian. Wuhan Publishing House, 2012.

Guo Shaofei. Blockchain Smart Contracts in Contract Law. *Oriental Law*, 2019, No. 3.

[US] Melanie Swan, *Blockchain*, translated by Han Feng, Gong Ming et al. New Star Press, 2016.

Li Xiangyu. *All Things Think Union: The Road to Digital Success*. Publishing House of Electronics Industry, 2018.

Liu Xizi. The Study on the Integrated Development of Blockchain and Artificial Intelligence Technology. *Cyberspace Security*, 2018, No. 11.

Wang Yanchuan. Blockchain: Laying the Foundation of Trust of Digital Society, *Guangming Daily*, November 17, 2019, p.7.

Li Yi. Four Characteristics of the Operating State of "Digital Society". *Study Times*, August 2, 2019, p. 8.

Zhang Yi, Zhu Yi. System Trust Based on Blockchain Technology: A Framework for Trust Decision Analysis. *E-government*, 2019, No.8.

Zhao Yongxin. Blockchain Trust Mechanism Promotes Inclusive Finance and Helps Solve Financing Problems of Micro, Small and Medium-sized Enterprises. *Securities Daily*, December 26, 2019, p. B1.

Wang Yukai. Smart Society Forces State Governance to Be Smart. *Information China*, 2018, No. 1.

Zheng Zhiming. It Is Urgent to Establish a Fundamental Platform of National Sovereignty Blockchain. *China Science Daily*, October 18, 2018, p.6.

Chapter 5
Community with a Shared Future for Humanity Based on Sovereignty Blockchain

In a risk society, unknown and unexpected consequences become the dominant force in history and society.

—Ulrich Baker, a German Sociologist.

We expect "Tech for Social Good" will become the universal value of the digital society, and this round of new technological revolution will truly bring about a better human civilization and a better future.

—Si Xiao, President of Tencent Research Institute.

If the gentleman worldwide pursues his conscience, then he can distinguish between right and wrong, identify likes and dislikes, regard others as his own, regard the nation as his family, and take all things worldwide as a whole. In this case, it is impossible for the world to lack governance.

—Wang Yangming, a Philosopher in the Ming Dynasty.

5.1 Human Destiny in Global Challenges

The world is changing steadily in a mighty way. The world today is experiencing earth-shaking changes, transformation, and adjustment. Multipolarity of the world, economic globalization, social informatization, and cultural diversity are developing in depth, Jinping (2017) and there have been conspicuous characteristics of the accelerated evolution propelled by the tremendous changes of the world, as well as more sources of global turmoil and risks.[1] "World economic growth needs new impetus, development needs to be more inclusive and balanced, the gap between rich and poor needs to be bridged, hot spots continue to be turbulent, and terrorism is rampant. The deficits of governance, trust, peace and development are serious challenges facing all

---

[1]Bao Yuanyuan. Changes in Global Governance System Provide New Opportunities for China. People's Daily Online. 2020. https://baijiahao.baidu.com/s?Id=1,655,573,043,212,048,646&wfr= spider&for=pc.

© Zhejiang University Press 2021
L. Yuming, *Sovereignty Blockchain 1.0*,
https://doi.org/10.1007/978-981-16-0757-8_5

mankind".[2] Atomic, biological, chemical, and digital weapons threaten the natural rights, life safety, and future development of all humans. Under the background of global integration, the embryo of a community with a shared future for humanity that "you have me and I have you" has formed. Global destiny is closely linked and shares weal and woe together.

5.1.1 Future Global Challenges

In the world today, the coupling among environmental system, social system, and economic system is gradually decreasing, and the contradictions are more prominent. Nuclear war, cyber war, financial war, biological war, and non-sovereignty forces together constitute the global crisis in which mankind is plunged. Under the impact of multiple challenges, the stability and security of human ecosystem are fundamentally changing, the world order reconstructed, and human civilization in crisis.

Threat of Nuclear War. Since the World War I, major nations worldwide have rushed to lay out their nuclear strategies and seek to establish a nuclear deterrent with both offensive and defensive capabilities. About 13 nations worldwide have developed nuclear weapons, and the threat of nuclear war is just around the corner. On August 6, 1945, the United States dropped an atomic bomb on Hiroshima, Japan, which was the first time in human history that nuclear weapons were used in war. On that day, more than 88,000 people were killed, 51,000 injured and missing. Of the 76,000 buildings in the city, 48,000 were completely destroyed and 22,000 were severely damaged.[3] In 1962, the Cuban missile crisis officially kicked off the threat of nuclear war. Experts and the public were worried human wisdom is not enough to avert this catastrophe, and it is only a matter of time before nuclear war breaks out (Noah Harari 2018). Facing the survival of mankind, China, the United States, the Soviet Union, and Europe united to establish an international alliance, which changed the geopolitics that had lasted for thousands of years and finally averted the catastrophe that might destroy mankind from happening. Until around 1970, the Soviet Union, the United States, Britain, and other nations launched new nuclear competition and nuclear games. New destructive nuclear weapons developed toward miniaturization, diversification, and portability, and the international nuclear rules that had been maintained for decades were broken. In 1986, the "Chernobyl Incident" broke out, "30 people died instantly and more than 8 tons of strong radiation leaked. More than 60,000 square kilometers of land around the nuclear power plant were heavily polluted and over 3.2 million people were exposed to nuclear radiation, leading

---

[2]Wu Yuzhen. *A Bigger Circle of Friends with More Consistent Ideas.* Xinmin Evening News, May 15, 2017, p. A3.

[3]Zhang Shuyan, Wang Jiawei. The United States dropped an atomic bomb on Hiroshima, Japan. August 6, 1945. People's Daily Online. 2013. https://history.people.com.cn/n/2013/0806/c364284-22457437.html.

to the greatest calamity in the human history of peaceful use of nuclear energy" (Hongbo 2017). As the global nuclear safety becomes increasingly prominent, the Iranian nuclear issue and the North Korean nuclear issue have become two major issues perplexing the international community today. Some nations have illegally developed nuclear weapons and repeatedly failed to comply with or even violated the provisions of relevant documents such as *the Treaty on the Non-Proliferation of Nuclear Weapons* (Yiming 2006). The hot conflicts between Iraq and the United States greatly increased the probability of restarting 20% enriched uranium activities involving weapons-grade nuclear technology and nuclear weapons. "Prevention of nuclear war" has become an inevitable prerequisite for human survival. Once vicious nuclear competition permeates the whole world, it will trigger the World War III marked by nuclear weapons and open the "door of disaster" for humans.

Threat of Cyber War. With the rise of network information technology in the second half of twentieth century, the Internet has built a bridge for international exchanges. Meanwhile, cyber warfare with hidden dangers of "Chekhov's Law" has emerged, and cyberspace has been an important battlefield in the future. The United States was the first nation to set up a cyber force, formulate a cyber warfare plan, and introduce the concept of deterrence into cyberspace (Yuqing and Yanli 2014). In June 2013, Snowden exposed the secret implementation of the U.S. government's electronic surveillance program code named "US-984XN" which attacked the global network 61,000 times. This was the famous "Prism Incident", a major event that shocked global cyberspace security and opened a new era of cyberspace warfare (Xingdong 2014). In May 2017, WannaCry blackmail virus spread all over the world, attacking nearly 300,000 devices in more than 150 nations (regions) including the United States, Britain, and China, Xiao and Junqi (2017) causing nearly 8 billion US dollars in economic losses. In order to safeguard national sovereignty, security, and interests, nations all over the world have deployed cyberspace security strategies one after another. Russia has completed the test of "Sovereignty Internet" and realized the independent operation of "Regional Internet" within its territory after the "network disconnection" operation. The nature of cyberspace characterizes cyber warfare by low threshold, vague boundaries, duality, difficult command, and management. When cyber forces carry out cyberspace missions, "Trojan horse programs" may spread to global cyberspace, non-target cyberspace may launch attacks due to self-defense, cyber forces may lose control, etc. When hackers or hacker organizations conduct network attacks, the target-free attack mode adopted makes it possible for any computer worldwide to become its target… All kinds of factors that can trigger network warfare are threatening the digital fate of humans. Cyber warfare has been truly displayed before humans. It has brought new means to destroy humans, whose threat is more than that of nuclear bombs, and it may put humans in a new out-of-order world.

Threat of Financial War. Financial security is related to economic progress and world peace, and is an important issue in global governance. Preventing and resolving financial risks, especially preventing systemic financial risks, is the fundamental financial task.[4] With deepening globalization and developing FinTech in recent years, the international financial system has closely linked the economic lifelines worldwide, forming an "economic community with a shared future" sharing weal and woe. Meanwhile, the competing, fighting, and crowding out in the financial field that nations are involved in are no less than any form of war in terms of scale, cost, benefit, and other aspects. Financial war has become a completely new form of demilitarized stealth warfare. For example, the Dutch tulip crisis in 1637, the British South Sea Bubble in 1720, the US financial panic in 1837, the US banking crisis in 1907, the US stock market crash in 1929, the Black Monday that swept the global stock market in 1987, the Mexican financial crisis in 1994–1995, the Asian financial crisis in 1997, the global financial crisis in 2008, and the Sino-US trade war in 2019. Every outbreak of financial crisis has brought serious chaos to social economic operation and is likely to collapse world economic system. The international financial game is a "battlefield without smoke of gunpowder", which is essentially a dispute over the dominance of global governance system. The most effective way in modern society to quickly defeat a big nation is not war, but finance. It can be said the fastest way to destroy a superpower is a financial war.

Threat of Biological War. Humans are facing multiple health challenges such as cancer, Ebola, AIDS, and coronavirus[5] and multiple social risks such as gene editing alienation. Bacteria, viruses, and toxins make people, animals, and plants sick or dead. Once used in modern warfare, "biological and chemical weapons" of mass destruction will surely become the greatest biological and chemical threat to human health and life safety. According to textual research, Japan started 30-year-long criminal bacteria experiments during the World War I until it was defeated in World War II in 1945 (Xiaoyan 2014). As history steers today, advanced life technology and medical technology are gradually eliminating diseases, but meanwhile they have also brought new survival challenges to humans. "Gene editing is a technology that can accurately modify the genome. It can carry out targeted knockout, knock in, simultaneous mutation at multiple sites and deletion of small fragments of genes of interest" (Yunling 2019). In 2017, a team headed by He Jiankui injected gene editing reagents into the fertilized eggs of six couples under the conditions that the law did not allow, ethics did not support and risks were uncontrollable, resulting in the birth of many gene-edited babies, seriously disrupting medical management order (Pan

---

[4]Xinhua News Agency. Xi Jinping: Deepening financial supply-side structural reform to enhance financial services for the real economy. Xinhuanet. 2019. https://www.xinhuanet.com/politics/lea ders/2019-02/23/c_1124153936.htm.

[5]Coronavirus is a kind of pathogen that mainly causes respiratory tract and intestinal diseases. The surface of this kind of virus is covered by many regularly arranged protrusions, and the whole virus is like a crown of emperors, hence the name "coronavirus". In addition to humans, coronavirus can also infect many mammals such as pigs, cattle, cats, dogs, minks, camels, bats, mice, hedgehogs, and many birds. So far, there are six known human coronaviruses. Four of them, more common in the population, have low pathogenicity. Generally, they only cause mild respiratory symptoms similar to the common cold. The other two coronaviruses, severe acute respiratory syndrome coronavirus and Middle East respiratory syndrome coronavirus which we call SARS and MERS, can cause serious respiratory diseases.

et al. 2019). This is not only a blatant challenge to all humans, but also turns a blind eye to the law, and their behavior was finally punished by the law.[6] Gene editing technology is originally designed to eliminate diseases, not to enhance certain functions of the human body system. Its abuse will destroy the natural evolution of human body system, subvert traditional morality and values, cause fatal blows to natural persons, and bring havoc to humans.

Threat from Non-sovereignty Power. The interconnectivity of nations worldwide is bound to usher in a Community with a Shared Future for Humanity. Concentration is a paradigm centralization dilemma in which wealth, belief, security, and other factors gather in human interaction (Goldin and Kutarna 2017). It has given birth to much non-sovereignty power. Hong Kong experienced troubles in 2019, afflicted by economic downturn, and traffic paralysis. According to statistics, the paralysis of the Hong Kong International Airport for 1 day caused a loss of nearly 210,000 passenger volumes, 13,863 tons of cargo transport, and nearly 10.2 billion Hong Kong dollars in air cargo value, harming the livelihood of over 800,000 Hong Kong people.[7] From June to September 2019, the economic loss caused by the reduction of tourists alone amounted to 18.5 billion Hong Kong dollars, with heavy losses in service industry, tourism, finance, and other fields. It can be seen that non-sovereignty power can affect the security of sovereignty nations by containing and controlling regional stability, thus undermining the international order for the long-term stability of human society although it cannot dominate international integration or control globalization.

5.1.2 Risks Behind "Grey Rhino" and "Black Swan"

The prosperous world linked by complex human relations, diverse financial systems, and rich scientific systems is not only the cornerstone for human civilization evolution, but the "hotbed" of various risks. The "Grey Rhino" and "Black Swan", two major types of risks that the world is most concerned about today, are only walking slowly but not jumping, so humans must take conscious precautions against them in order to "fight back".

---

[6]On December 30, 2019, the case of "Gene Editing Baby" was publicly pronounced in the first instance by Shenzhen Nanshan District People's Court. He Jiankui, Zhang Renli, and Qin Jinzhou area were investigated for criminal responsibility according to law for jointly and illegally carrying out human embryo gene editing and reproductive medical activities for reproductive purposes, committing illegal medical practice. According to the defendant's criminal facts, nature, circumstances, and degree of harm to the society, the defendant He Jiankui was sentenced to fixed-term imprisonment of 3 years and a fine of 3 million yuan; Zhang Renli, fixed-term imprisonment of 2 years and a fine of 1 million yuan; and Qin Jinzhou, fixed-term imprisonment of 1 year and 6 months, suspended for 2 years and a fine of 500,000 yuan.

[7]Zhu Yanjing. Hong Kong Airport Paralyzed; Passengers Suffered; the Livelihood of More than 800,000 Hong Kong People Affected. China News. 2019. https://www.chinanews.com/ga/2019/08-13/8924885.Shtml.

5.1.2.1 "Grey Rhino" and "Black Swan"

"Grey Rhino". In *Grey Rhino: How to Deal with Probability Crisis*, Michel Walker put forward the concept of "grey rhino", which is used to describe the potential risks with high probability and great influence and are taken for granted by people, such as real estate, financial crisis, and resource competition. Grey rhino is characterized by large size, stupid actions, and slow response. Once it launches an attack, the risk will soar. At that time, the options before humans are no longer good or bad, but terrible, more terrible, or even beyond redemption (Wucker 2017). The UN report pointed out that "In the past 50 years or so, 507 of the 1831 cases caused by water disputes have been of a conflict nature and 37 have been of a violent nature, while 21 of these 37 have turned into military conflicts. According to relevant agencies, by 2050, the number of nations threatened by water shortage will increase to 54, and the affected population will account for 40% of the global population, reaching 4 billion people" (Zhifei 2013). In fact, if humans carefully review the crisis after it happened, they will find that the clues before the major crisis are actually excellent "escape" opportunities (Wei 2017).

"Black Swan". Taleb's *The Black Swan: The Impact of The Highly Improbable* endows "Black Swan" with a new connotation, that is, extremely unpredictable, unusually accidental, or unexpected events, Song and Yaqian (2017) such as the "9/11" terrorist attack, Britain's Brexit, Trump's victory in the election, Italy's failure in the constitutional amendment referendum, and a large-scale virus pandemic. There were countless large-scale virus outbreaks in human history: the first discovery of HIV in the United States in 1983; the outbreak of SARS in China in 2003; H5N1 "bird flu" in Southeast Asia in 2005; influenza a (H1N1) in Mexico in 2009; "Middle East Respiratory Syndrome" first discovered in Saudi Arabia in 2012; "Ebola virus" in Guinea, Africa in 2014; and "Zika virus" in Chile, South America in 2014. Every large-scale virus outbreak brings deaths to humans. COVID-19 swept across the nation and the world in less than 2 months. Wuhan shut down its city and China launched a first-level response. The World Health Organization announced the epidemic constituted a "public health emergency of international concern" (PHEIC). This shows that once the "Black Swan" strikes, it will push the world to the extreme edge of a global disaster.

To be Keenly Aware of "Black Swan" and "Grey Rhino". On July 17, 2017, the first working day after the National Financial Work Conference was held, the front-page editorial *Effective Prevention of Financial Risks* of *People's Daily* mentioned that it is necessary to be keenly aware of potential dangers, including "black swan" and "grey rhino" in order to prevent and resolve financial risks. We cannot take all kinds of risks lightly, let alone turn a deaf ear to (Xuebin 2018). "Grey Rhino" and "Black Swan" continue to be in human vision through authoritative channels, alerting humans to nip in the bud. Although the crisis brought by "grey rhino" is often extremely destructive, it is predictable, perceivable, and preventable, but humans often adopt "ostrich tactics" passively (Jie and Yiyun 2017). Michelle Walker said that many crises usually had obvious signs before breaking out, but humans always looked at these signs with fluke and even arrogance until they broke out. "Black

Swan refers to unpredictable major and rare events. It often brings unpredictable major impacts. However, turning a blind eye, humans are used to explaining with their limited experience and fragile beliefs. Eventually, they are defeated by reality" (Nicholas Taleb 2011). All risks are likely to impact the bottom line of human destiny and destroy the achievements of human development. The best plan is to transcend, change, and renew traditional ideas.

5.1.2.2 Approaching Global Risks

As the world is increasingly diversified and interconnected, incremental changes have been replaced by instability, threshold effects, and chain damage of feedback loops. Global risks with high probability and strong destructiveness are pushing humans to the "sea of death". "Grey Rhino" is running at full speed toward humans, such as failure of climate change and adjustment measures, water resources crisis, cyber-attacks, natural disasters, and failure of key information infrastructure. Humans are facing risks behind climate deterioration, sci-tech abuse, and depletion of resources. "Black Swan" is also coming quietly, such as weapons of mass destruction, man-made environmental disasters, extreme weather events, biodiversity loss and ecosystem collapse, and assets bubbles in major economies. Humans are facing crises arising from the proliferation of weapons, economic collapse, ecological imbalance, and other related factors (World Economic Forum 2019). The human society is already in jeopardy. More and more people come to realize that they need to make greater efforts and changes to find ways to all the challenges.

Looking into the future, the two major crises will inevitably catch our attention: the first is global warming. At present, the content of carbon dioxide in the air is rising at an unprecedented rate. Ice melting in the Arctic will cause sea level to rise by more than 1 m by the end of twenty-first century. Storm surges and floods will plunge 1 billion people worldwide into crises and nearly 300 million homeless. Steady rise in average global temperature and further melting of polar glaciers will lead to changes in ocean currents or inundation of lower lying coastal areas, or aggravate desertification in areas that can still maintain farming nowadays (Ferguson 2012). The second is natural disasters. With the narrowing gap between rich and poor nations, mankind is facing extreme and catastrophic climate changes. On February 12, 2020, Australian jungle fire lasting 210 days was finally put out and the "scar on the earth" finally stopped expanding. The fire killed at least 33 people, about 1 billion wild animals, and burned more than 2,500 houses and 11.7 million hectares of land.[8] There is no doubt the environment on which humans depend is already overwhelmed and global environmental risks are increasing day by day.

Future risks happen one after another, pushing humans to the abyss of extinction. The first risk is meteorological control tools. The unilateral use of radical geo-engineering technology will cause climate chaos and aggravate geopolitical tensions.

---

[8]Guo Xinwei. Having burned for 210 days! The fire in NSW was finally extinguished. China News. 2020. https://www.chinanews.com/gj/2020/02-12/9088580.Shtml.

The second risk is the gap between urban and rural areas. The widening gap has intensified the polarization between nations and regions. When the gap approaches a critical value, local nativism and violent conflicts may occur. The third risk is exhausted natural resources. When the resources that nature produces cannot satisfy humans, the battle for resources will be staged and society will surely be in turmoil. The fourth risk is the Space Battle. Nations worldwide are scrambling to lay out satellite systems in space and seize dominance in space. Space debris is rushing to the earth at the speed of bullets. The fifth risk is the loss of human rights. When a superpower is in control and domestic division is intensifying, the government tends to sacrifice individual interests for collective stability. Humans have no human rights at all (World Economic Forum 2019).

5.1.2.3 Reflections Behind Global Risks

Global issues belong to three major fields of relations. First, imbalance between humans and nature; second, imbalance between humans and society; and third, imbalance between humans and themselves (human's grasp of the relationship as the subject) (Zhiping 2003). The imbalance between humans and nature often brings about the imbalance between humans and society, and between people and themselves, which is also the prominent feature of current global risks. Global population has been soaring since twentieth century. Global population is expected to reach 8.5 billion in 2030, 9.7 billion in 2050, and 10.9 billion in 2100.[9] When it becomes larger, global population will definitely intensify the contradictions and conflicts in social economy, environmental resources, human relations, civilized politics, etc. Then, the imbalance in the three major areas will be more complex and interconnected. The global crisis facing humans at this stage is the product of history, while the risks facing humans in the future depend on the activities, behavior patterns, and value choices of contemporary humans. Only by reviewing the internal ideological and cultural patterns of civilization, science and technology, politics, and others in human civilization can we truly prevent and eliminate global crisis and bring the world back to a safe, stable, bright, and brilliant future.

"Human society is a complex society full of nonlinearity, uncertainty, vulnerability and risks. With a globalized human society, modernization, steady progress of sci-tech innovation and profound changes in international politics, human society has undergone profound systematic structural transformation into a highly uncertain and complex global risk society".[10] In the past centuries, "Grey Rhino" and "Black Swan" were born in the choice of humans, and they may be the last blow to humans. Therefore, no matter how humans make choices about drawing on advantages and

---

[9]Department of Economic and Social Affairs. "World population prospects 2019: Highlights". United Nations. 2019. https://www.un.org/development/desa/publications/world-population-prospects-2019-highlights.html.

[10]Fan Ruguo. "World Risk Society" Governance: The Paradigm of Complexity and Chinese Participation. *Social Sciences in China Press*, 2017, No. 2, p. 65.

avoiding disadvantages, they should visualize the disadvantages behind their choices. However, the reality is that humans are constantly warned, nerves tightened, risks uncontrolled, crises coming quickly, and human society is in a cycle of anxiety and helplessness. In the face of so many risks, humans should frankly admit their own defects to avoid collapse.

This is an era of rapid evolution of technologies and resources, and it is also an era of extreme insecurity. The contract linking humans, nature, and science is disintegrating. Today, humans are enjoying the highest living standard in history, advanced science and technology, and abundant financial resources. Humans should use these resources rationally, adhere to the principle of sustainable development, and move forward toward a fairer and more inclusive future. However, due to the lack of sufficient motivation and in-depth cooperation in the changes, humans may push the world to the brink of system collapse.[11] While the international landscape is undergoing complex evolution today, when closure and openness, unilateralism and multilateralism are intertwined, the goal sought by humans is the future of a common destiny. The future of human destiny is filled with unknowns. The future may fall into a winter of disappointment, or brightness with threats and challenges sleeping forever and disasters not happening.

5.1.3 The Gospel or Disaster of Artificial Intelligence

Humans, the only essential element in the global ecosystem with their own wisdom, are driving global changes. Artificial intelligence is the product of humans in their adaptation to nature, such as improving living environment, optimizing quality of life, and improving production efficiency. It is an important technical form to satisfy human needs. The appearance of Homo sapiens rewrites natural selection and traditional rule systems, shakes the boundaries between ecosystems on the earth, and extends the life form from the organic field to the inorganic field.[12]

5.1.3.1 Four Waves of Artificial Intelligence

First Wave: Computational Intelligence. From 1950 to 1980, the rapid progress of natural language processing capability of computers achieved computational intelligence and fast storage, and tackled many problems which seem to be addressed only by human wisdom, e.g., maze, IQ games, and chess (Yutaka 2016). As the most famous research result at this stage, Turing test isolates people from machines

---

[11] World Economic Forum. "The global risks report 2018 (13th edition)". World Economic Forum. 2018, http://www3.weforum.org/docs/WEF_GRR18_Report.Pdf.

[12] [Israel] Yuval Noah Harari. *Homo Deus: A Brief History of Tomorrow*, translated by Lin Junhong. China CITIC Press, 2017, p. 67.

and becomes an important means to test artificial intelligence. Although computational intelligence at this time can only solve some basic problems, it cannot prevent the emergence of sci-tech products. ARPANET, the world's first prototype of computer network with substantial significance, has achieved digital data transmission in computer history for the first time. The economic value created by computational intelligence is limited to the high-tech industry and the digital world (Kaifu 2018). Therefore, humans realize it is still extremely difficult to use artificial intelligence to solve practical problems.

Second Wave: Perceptive Intelligence. The integration of traditional industries and artificial intelligence in the mid- and late 1990s was the most remarkable achievement in this stage. Artificial intelligence, with knowledge as its support, needs as its soul, rules as its skeleton, steadily iterates and updates and realized perceptive intelligence such as vision, hearing, and touch. Automatic driving applies AI technology in gauging the distance, orientation, and speed between external people, vehicles, and objects through sensors, and transmits the information to an intelligent perception module for calculation and processing. Baidu and Google's perceptive intelligence technology led the second trend of artificial intelligence. The market value of Google Waymo[13] was estimated at 175 billion US dollars by Morgan Stanley. Taking this as a reference, Baidu Apollo's market value will be 30 billion to 50 billion US dollars. Perceptive intelligence affects practical problems, such as smart application, terminal replacement, and financial economy to a certain extent, but there is still a "bottleneck of knowledge acquisition".

Third Wave: Cognitive Intelligence. The twenty-first century is rapidly changing. With abundant resources and the combat capability of large armies, technology giants have taken the lead and brought the third wave to the world with the introduction of intelligent products such as Watson System and AlphaGo System (Lei 2017). At present, the biggest gap between machines and humans lies in cognitive intelligence, which is also an area where major technology giants are urgently looking for breakthroughs (Hui 2016). Symbolism of knowledge graph and connectionism of deep learning make cognitive intelligence possible. They transform the real world into a quantifiable, analyzable, and computable digital world, such as Xiaomi's "Classmate AI", Alibaba's "City Brain", and the face recognition of Face ++. They show that cognitive intelligence can be realized, applied and developed. Once cognitive intelligence is realized, the familiar life pattern and social pattern of humans be changed, and the real physical world be digitized.

Fourth Wave: Super Intelligence. "Humans are facing a quantum transition, the strongest social change and creative reorganization in history" (Toffler 2018). In October 2019, Google successfully demonstrated a 53-qubit processor, allowing

---

[13] Waymo was originally a self-driving car program launched by Google in 2009, and then became independent from Google and a subsidiary of Alphabet (Google's parent company) in December 2016.

quantum systems to take about 200 s to complete tasks that traditional supercomputers would take 10,000 years to complete.[14] It can be predicted that strong artificial intelligence which is equivalent to human intelligence and ability, and super artificial intelligence which surpasses humans hundreds of millions of times in all aspects, will appear one after another. The future is the era of super intelligence, AI + QI = SI (Artificial Intelligence + Quantum intelligence = Super Intelligence). The overlapping of quantum intelligence and artificial intelligence is the technical guarantee for humans to move toward digital civilization. In the era of super intelligence, the structure of the entire human ecosystem will change. Humans have been caught in the wave of change while they have not yet clearly understood this fact.

5.1.3.2 The Advent of Gospel: Artificial Intelligence in Progress

From the perspective of traditional thinking pattern, there are differences in resource allocation, moral level, and economic income. However, with the advent of artificial intelligence, everything has returned to a state of remixing. After experiencing the Industrial Revolution, World Wars, and numerous changes and turbulence, human society comes to realize that individual freedom, stability, and development cannot be separated from the efficiency and justice of the government and society (Yanhong 2017). Looking at the artificial intelligence today, the gap between China and Western developed countries is narrowing in sci-tech terms. We see a technological trend that is about to impact global economy and make the geopolitical balance toward China (Kaifu 2018). Regarding the research and application of artificial intelligence, traditional American companies have done well in the commercial field, while China has performed well in the livelihood of the people. Although the two modes and strategies are different, time will definitely tell which is better. The era of digital civilization is fair, equitable, and open. All nations worldwide have the rights to compete equally in the field of artificial intelligence.

Food, clothing, housing and transportation, complex social relations, and abundant natural resources are all digitized, and the amount of data generated will gradually exceed the amount of data that can be stored on the existing Internet. Data intelligence is the key to this problem. Through technologies such as big data engine, in-depth learning and machine learning, data intelligence cleans, classifies, and calculates massive data, configures its transmission, application, and storage in an intelligent way, and taps into the maximum value contained in the data. Instead of being confined by huge amount of data or physical equipment, AI will work with humans to address global issues, such as cancer, climate change, energy, economics, chemistry, physics, and many others. The breakthrough of AI technology will improve productivity and production efficiency and accelerate the arrival of a new round of sci-tech revolution.

Artificial intelligence has been applied to all fields of society, bringing another opportunity for social restructuring (Zhiyong 2016). It provides strong support for

---

[14]Cheng Lan. Google Researchers Announce Successful Demonstration of "Quantum Hegemony". Xinhuanet. 2019. https://www.xinhuanet.com/2019-10/23/c_1125143815.Htm.

human society in the era of digital civilization, making impossible things possible. Intelligent travel, smart homes, intelligent wearables, intelligent medical treatment, intelligent courts, and so on have been applied to human society, and the road of social intelligence has been paved. In the field of search, artificial intelligence is not dependent on a fixed knowledge base or is limited to general retrieval. It associates available information from many data sources, (Hendler and Mulvehill 2018) and ranks the information according to the needs of humans in order to enable humans to obtain the most useful information in the shortest possible time. In the field of transportation, an experiment conducted by Google shows that if 90% of cars are driverless, the number of car accidents will drop from 6 million to 1.3 million, and the death toll will be greatly reduced. Meanwhile, driverless cars can also avoid artificial traffic jams.[15] Artificial intelligence is sending a steady stream of "Force" to human society.

5.1.3.3 Singularity Approaching: Hidden Worries in Artificial Intelligence

As artificial intelligence technology becomes more advanced, it increasingly replaces a large amount of traditional manual labor of humans and it adapts very quickly (Kurzweil 2011). A report released by Price Waterhouse Coopers predicts that by the early 2030s, 38% of jobs in the United States will be threatened by automation, compared with 30% in Britain, 35% in Germany, and 21% in Japan.[16] More and more jobs are being replaced by machines, and the jobs that have not been replaced will be filled by humans with smarter brains and stronger physical strength. This modern society constructed by humans and nature is gradually becoming a future society based on the "Golden Triangle" framework of humans, nature, and artificial intelligence. This is incompatible with the existing social structure and its internal distribution order, which potentially means that the existing social system needs to be upgraded, otherwise it will bring about violent conflicts within humans.

It will be difficult to avert the severe impact of AI in the face of AI. AI is not part of natural order, but the product of human creativity, which may aggravate human conditions (Walsh 2018). The legal system is an important benchmark to maintain social stability and security. However, the existing legal system is helpless to artificial intelligence. As traditional rules gradually fail, the risks of institutional mechanisms will follow. Artificial intelligence not only challenges current legal rules, ethical rules, and social order, but highlights the defects of current legal system, especially when current laws still have gaps in the field of artificial intelligence. Therefore, humans not only need to give legal care to artificial intelligence, but also need to set restrictions and responsibilities for its development, so as to ensure that it will benefit mankind rather than destroy mankind (Yanchao 2019).

---

[15]Wang Ruihong. Artificial Intelligence Welcomes Promising Development Chances. *Times Finance*, 2017, No. 16, p. 42.

[16]Ibid.

Ray Kurzweil, Google's CTO and futurologist, put forward the famous "Singularity Theory": "Technology will achieve explosive growth at a certain point in the future and break through a critical point, which is the 'singularity'. At that time, human civilization will be completely replaced by artificial intelligence" (Jinchang 2017). The arrival of singularity crises, e.g., normative singularity, classical theoretical singularity, economic singularity, social form singularity, and technological singularity (Zhangcheng 2018) may plunge human fate into great threat. Humans become the dominant species on the earth, to a large extent, because human intelligence surpasses other species. Artificial intelligence was originally an external object conceived and created by humans. If the wisdom of "homo sapiens" surpasses that of humans, it indicates artificial intelligence will compete with humans for dominance. Then in the face of artificial intelligence, humans will seem to be incapable, and sometimes even confused, perplexed, and helpless (Caihong 2018).

5.2 Governance Technology and a Community with a Shared Future for Cyberspace

A new round of sci-tech revolution and industrial transformation at present is accelerating, and new technologies, applications, and business forms such as artificial intelligence, big data, and the Internet of Things are on the rise. The Internet has gained stronger momentum and broader space for development. It is the common responsibility of the international community to develop, utilize, and govern the Internet to bring greater benefits to humans. Under the international initiative of the Four Principles and Five Propositions,[17] we will make innovations in governance technology, promote the reform of global Internet governance system, realize the benign transformation of cyberspace from "technological governance" to "sovereignty governance", and build a Community with a Shared Future for Cyberspace featuring respect for cyber sovereignty.

---

[17]The Four Principles and the Five Propositions were first put forward by President Xi Jinping at the opening ceremony of the Second World Internet Conference on December 16, 2015. The "Four Principles" refer to the principles that should be upheld in reforming global Internet governance system: respecting cyber sovereignty, safeguarding peace and security, promoting openness and cooperation, and building good order. The Five Propositions were proposed by President Xi Jinping for building a Community with a Shared Future for Cyberspace. First, stepping up the development of global network infrastructure and connectivity. Second, building an online platform for cultural exchange and sharing, and promoting exchanges and mutual learning. Third, boosting innovative development of the cyber economy and common prosperity. Fourth, ensuring network security and promoting orderly development. Fifth, building an Internet governance system for equity and justice.

5.2.1 Governance Technology and Governance Modernization

The Fourth Plenary Session of 19th Central Committee of the Communist Party of China proposed that adhering to and improving the socialist system with Chinese characteristics and promoting the modernization of China's governance system and governance capacity is a major strategic task for the CPC. Governance technologies centering on digitalization, networking, and intelligentizing are emerging and continue to unleash governance effectiveness. Governance technologies are an essential support for national modernization, and an organizational method for datalization of power and data empowerment. The central issue it highlights is "datalization" governance. Innovating governance system, and improving governance approaches and governance level with governance technologies are major channels for the modernization of China's governance system and governance capacity and the key to shifting the "Governance of China" to the "Chinese Dream".

5.2.1.1 Governance Revolution Under the Complex Theory

Human political history is the process from "ruling", "managing" to "governing". The word "governance" is originated from Latin and Greek, with its meaning of "control, guide and manipulate", mainly used for management activities and political activities related to national public affairs (Lijuan 2012). Among the various definitions of governance, the one given by the United Nations Council on Global Governance is the most representative. According to its definition, "Governance is the sum of the many ways in which individuals and institutions, public or private, manage their common affairs. It is an ongoing process by which conflicting or different interests are reconciled and united in action is taken. It includes both formal institutions and rules that have the right to compel obedience and informal institutional arrangements that people agree to or believe are in their interests". This definition suggests that governance has four basic characteristics. First, governance is not a set of rules or an activity, but a process; second, the foundation of governance process is not control, but coordination; third, governance involves both public and private sectors; fourth, governance is not a formal system, but a continuous interaction. Obviously, governance goes beyond traditional bureaucratic system and democratic system. It regards the management of public affairs as a process of multi-subject participation and multi-responsibility sharing, as well as a process of multi-mechanism resonance and multi-resource integration (Yuming 2014).

The research and practice of governance theory emerged in Western nations in the late 1990s, when Western society developed it in response to market failure and government shortfalls. The West has a long history in trusting markets and governments, but neither is a panacea. It is impossible to find a solution to the "market failure" in the market system, so the government is pushed forward as a rectifier. However, the excessive involvement of the government in the economic field has weakened economic vitality, penetrated into social fields without any limits, reduced the space for people's free development, and led to the sharp decline of the social restriction function. It is in this context that governance, as a new way to allocate social resources, emerges as an effective way for the benign interaction between the government, the third sector and non-profit organizations and other social forces, and provides a new solution to the dual failure of the government and the market.

Governance theory is inseparable from the rise of complex theory. In the 1970s and 1980s, some paradigm crises in social sciences gave rise to complex scientific paradigms and laid a foundation for governance theory. At that time, the classical scientific paradigm fails to give a good description of the real world. With the development of information technology revolution and the formation of knowledge economy and circular economy, a post-modern "Complexity Science" was formed to study the complexity and nonlinearity of systems (Baobin et al. 2013) based on the theory of atomic structure, quantum mechanics, and relativity in "modern scientific revolution". Qian Xuesen, an advocate of Complexity Theory in China, defined the complexity of system science as "The 'complexity' we understand is actually the dynamics of open complex giant systems, or open complex giant systematics" (Xuesen 2007). The birth of Complexity Theory is a radical change of thinking pattern for the development of governance theory. Now it seems governance theory is a kind of complexity science paradigm, which seeks the benign interaction mechanism among society, government, and market, as well as a complex mechanism related to politics, economics, sociology, and urbanology. Governance theory is difficult to achieve perfection without complexity thought. Both complex theory and governance theory are based on the complex social system of post-industrial society.

5.2.1.2 Governance Technology to Modernize Governance

Management modernization is "Fifth Modernization" after modernization in industry, agriculture, national defense, and science and technology, and its essence is the modernization of system. In November 2013, the Third Plenary Session of 18th CPC Central Committee adopted the Decision of the *CPC Central Committee on Some Major Issues Concerning Deepening Reforms Comprehensively*, which stated "promoting modernization of China's governance system and governance capacity". This is the first time to link national governance system and capacity with modernization, focusing on modernization and taking modernization as the ultimate goal, revealing the close internal relationship between modernization and national governance. National governance is inseparable from modernization, and modernization constitutes the inherent meaning of national governance (Yaotong 2014). Six years later, in October 2019, the Fourth Plenary Session of 19th CPC Central Committee adopted *the Decision of the CPC Central Committee on Several Major Issues Concerning Upholding and Improving the Socialist System with Chinese Characteristics and Promoting the Modernization of China's Governance System and Governance Capacity*, in which some new ideas, technologies, and modes, including "Internet", "big data", "artificial intelligence", "digital government", "tech support", and "science and technology ethics", were written and they become a fresh power to modernize national governance system and governance capability. Starting from this milestone, China has stepped into a new era that governance technologies promote management modernization.

Governance technology will become a key force and play a bigger role in promoting the modernization of China's governance system and governance capacity, providing new support for the new requirements of national governance modernization with new technological means and operational mechanisms. Especially when the nation is in a critical moment, governance technology shows strong vitality and superiority in governance modernization by virtue of its unique system arrangements and technological advantages of "dual drive". "ABCDEFGHI" collaborative innovation is evolving into a driving force for the modernization of national governance system and governance capacity under the framework of governance science and technology. "ABCDEFGHI" refers to key technologies of artificial intelligence (AI), Blockchain, Cloud Computing, Big Data, Edge Computing, Federated Learning, 5th Generation, Smart Home, and Internet of Things. These emerging technologies are constantly converging, producing chain-like, interdisciplinary group innovation achievements. The parallel development of subversive and revolutionary innovations and iterative and progressive innovations is reconstructing the underlying infrastructure and operation logic of national governance.

The "soul" of governance technology is governance, while technology is only its "principle". Its core is to achieve "four transformations" via two-way linkage and multi-direction empowerment of "governance" and "science and technology". The first is the transformation from "governing people" and "governing objects" to "governing data". The modernization of national governance will be promoted through a five-in-one system of digital governance, including national digital, social digital, urban digital, economic digital, and cultural digital governances (Duan 2019). The second transformation is from the concept of "national management" to "national governance". It puts more emphasis on the flexibility, coordination, and communication of governance, which demonstrates the fairness and justice of the nation and the harmony and order of the society. As President Xi remarked, "The gap between governance and management is a reflection of systematic governance, law-based governance, source governance and comprehensive measures". The third is the structural transformation from "unitary subject" to "pluralistic subject". Governance is realized by the concerted efforts of the government, markets, social organizations, CPC committees, the NPC, the CPPCC, and other pluralistic subjects, rather than by a single force, and the consultation, co-governance, and sharing among governance subjects are highlighted. The fourth is a comprehensive transformation from "administration" to "politics, rule of law, rule of virtue, autonomy and wisdom". It aims to exercising their key role in promoting the modernization of national governance with strong political guidance, legal guarantee, moral education, autonomy, and intellectual support. It can be said that governance science and technology promote governance modernization, which on the surface is only a change of supporting elements, but behind it lies a well-read paper from vertical to delayering, from one way to a system, from command to the rule of law, from symptoms to root causes, and from unitary subject to diversified cooperation. And this well-read paper is the key force for China to stand out as a peaceful nation among the nations of the world.

5.2.1.3 International Order and Chinese Approach

"International order, different from international relations, is not simply made up of specific relations such as Sino-American relations, US-European relations, but an integral concept, an overall understanding of international relations, and a definition of the basic characteristics of international relations at a certain stage (Yugang 2014). Henry Kissinger put forward three valuable view in his *World Order*. First, there was never one order worldwide, but multiple ones based on religion, empires, or later, on sovereignty states. Second, each civilization has its own view of international order different from other civilizations. Therefore, as a civilization emerges and becomes dominant, its view of international order will inevitably influence the international order established by it. Third, the international order established by the West has been dominant since modern times, spreading from the West to the rest of the world. However, though it is a fact that the West dominates the world, it does not mean Western order is the only one. Different forms of regional order are emerging worldwide, exerting impact on the international order.[18]

Over the past 200 years, Western nations, America in particular, have played a leading role in constructing modern international order, so they have the right of definition and discourse about international order. "The United States has the right to define, which means whatever actions it takes globally, it can always justify to its people or the international community. There is no doubt the right to define has profound moral implications that attest to the 'justice' of American behavior and even war" (Yongnian 2019). "The US-led international order has three pillars. First, American values, also seen as 'Western values'; second, the military alliance of the United States constitutes the security cornerstone for it to play a 'leadership' role worldwide; third, international institutions, including the United Nations" (Ying 2016). The American-style international order has its historical origins in international politics and also functions in the modern world. However, worldwide of deepening economic globalization and fragmented international politics today, the world is experiencing tremendous changes unseen in a century, and international order is facing adjustment and reorganization unforeseen in history. The world has changed and is destined not to return to its original point. The US-led international order is increasingly difficult to provide comprehensive and effective solutions to international problems.

The biggest international political change at the beginning of twenty-first century is the sustained development of China. After more than four decades of reform and opening up, China has grown from a marginal role in the international community

---

[18]Zheng Yongnian. The stage of passive response has passed. Experience shows that passive response, no matter how good it is, is far from enough—an effective response to the US right to define. *Beijing Daily*, September 2, 2019, p.16; [US] Henry Kissinger. *World Order,* translated by Hu Liping, Lin Hua, Cao Aiju. China CITIC Press, 2015, Preface.

to a prominent role in global economic, political, and security fields (Kejin 2017). Its global impact and international discourse have never been greater. In this international context, China's concept of international order has become the focus of the international community (Table 5.1), and the "Governance of China" has become the "Oriental Wisdom" discussed by nations worldwide. Peace, development, cooperation, and win–win results are the key words in China's concept on international order. Chinese leaders have repeatedly stated that China will unswervingly support international order and system with the purposes and principles stated in the *UN Charter* as its core, and will be a builder of world peace, contributor to global development, and defender of international order. China has a sense of belonging to the existing international order. China is one of its founders, beneficiaries, contributors, and reform participants. The "Belt and Road" Initiative, the Asian Infrastructure Investment Bank, and a Community with a Shared Future for Humanity are important new public goods that China has provided to the world. Since its birth, the "Governance of China" has never been exclusive but inclusive. It has never been a winner-takes-all approach but win–win cooperation. It has never been hegemonism but "doing after discussing". It can be said that the "Governance of China" is not only a vivid interpretation of China's Concept of international order, but also a "Chinese solution" that China's peaceful development in twenty-first century contributes to the world.

5.2.2 Three Pillars of Governance Technology

Governance technology, a major innovation in the new-era "Governance + Sci-tech", is the great practice that science and technology empower management. An important way to realize governance modernization is to innovate governance system by governance technology, upgrade governance methods, and governance level. Block data, data rights law, and sovereignty blockchain constitute the "three pillars" of governance technology. Among them, block data is the data philosophy centralized at humans, the data rights law is the new order of humans toward digital civilization, and the sovereignty blockchain is the rule of technology governance under law. The interaction of "three pillars" forms a unified organism and constitutes the solution to the global governance system of the Internet and an important turning point in the era of artificial intelligence.

5.2.2.1 Block Data

A new round of sci-tech revolution and industrial transformation currently is at an important stage of convergence. With the interconnectivity of information technology and human production and life, the rapid popularization of the Internet and the growth and concentration of global data have exerted a significant impact on economic development, national governance, and people's life. We have entered a

Table 5.1 Evolution of China's international order concept

Stage	Year	Overview
Embryonic stage	1949	The founding of the People's Republic of China ushered in a new era of independent participation in international affairs and integration into the international order. China proposed such diplomatic ideas, concepts, and strategies as "one-sided" strategy, the Five Principles of Peaceful Coexistence, the idea of "One Line, One Large Area", and the theory of "Division of Three Worlds". The cognition of the international order was embedded in these concepts, and they were adjusting and changing to China's specific practice
	1974	At the sixth special session of General Assembly, Deng Xiaoping reaffirmed Mao Zedong's "Division of Three Worlds" theory, attacking the two superpowers: America and the Soviet Union as well as the old order based on colonialism, imperialism, and hegemony. He consented and supported the appeal and reform suggestions of the third world nations to change highly unequal international economic relations, and proposed the political and economic relations between nations should be based on the Five Principles of Peaceful Coexistence
Exploration stage	1978	China entered a new period of reform internally and opening up to the outside world, and achieved a great turning point in its development. Based on the review of the topics of the times and the international environment, the focus on China's concept of international order changed to the establishment of a new international political and economic order for maintaining an international peaceful environment conducive to China's economic construction. Chinese leaders made an important judgment on the two major themes of the times: "peace and development", and began to pursue an independent non-alignment policy
	1989	The report on the Government Work at the Second Session of 7th Seventh NPC put forward the official proposition of establishing a new international political and economic order in the initiative of the Chinese government
	1991	The Fourth session of 7th NPC listed the establishment of a new international order as an important part of China's foreign policy. Facing the complex international situation, China advocated "respecting the diversity of world civilizations, ensuring harmonious coexistence and mutual respect among nations", "promoting democracy in international relations, and pooling the strength of people of all nations to solve prominent problems". Under the guidance of a series of scientific judgments, China carried out multi-directional and multi-channel exchanges and cooperation with others worldwide, and came to develop in concrete practice a concept of international order that is rich in content and interconnected by all parties
Perfect stage	2005	China put forward the idea of building a "harmonious world" featuring lasting peace and common prosperity
	2007	The report of 17th National Congress of the CPC stated China "will continue to take an active part in multilateral affairs, assume corresponding international obligations and play a constructive role in promoting a more just and equitable international order"

(continued)

Table 5.1 (continued)

Stage	Year	Overview
	2011	"Nations with different systems, types and stages of development are interdependent and their interests are so intertwined that they form a community with a shared future", the white paper on China's Peaceful Development proposed. Since 18th National Congress of the CPC, China's leaders have advocated the idea of A Community with a Shared Future for Humanity, maintaining the international order and international system with the purposes and principles of the UN *charter* as the core, establishing new international relations with win–win cooperation as the core, building A Community with a Shared Future for Humanity as a key proposition of China in bilateral and multilateral relations and international order. As a firm defender, builder, and contributor of the international order, China set its goal in promoting the necessary reforms and improvement of international order and system to make them more just and reasonable

Source Dong He, Yuan Zhengqing, China's View on International Order: Formation and Core, *Teaching and Research*, 2016, No. 7

new information-intensive stage marked by big data. Humans will truly step into the era of big data with block data as the symbol. Block data is the advanced form of big data development, the core value of big data fusion, and the solution in the era of big data. By converging scattered point data and segmented bar data on a specific platform, block data exerts steady aggregation effects with them. Through multi-dimensional data fusion and correlation analysis, the aggregation effects can conduct research and predictions on the future tendency in a quicker, more comprehensive, more accurate, and more effective way so as to reveal the essence and laws of objects and promote the evolution of order and civilization.

Not only will block data bring us new knowledge, new technologies, and new perspectives, it will also revolutionize our worldviews, values, and methodologies. Block data, as a new philosophical thinking, reconstructs social structure, economic function, organizational form, and value world, and redefines future human social composition with natural persons, robots, and gene-edited men as the subjects. Its core philosophy is to advocate the people-oriented spirit of altruism. Block data, as a theoretical innovation on the basis of technological progress, reconstructs current economic system and social system and brings a new scientific revolution and social revolution, which is a paradigm of data sociology with the origins in humans. It analyzes human behavior, understands the rule of humans, and predicts human future with data technology rather than human thinking. It has profoundly changed the current modes of ethical thinking, resource allocation, value creation, rights distribution, and legal adjustment. Block data, like the human base in the data world, is the starting point of human cognition of this new world. More and more "bases" established in the data world turn out to be one piece and a new world, which means the birth of a new civilization—arrival of the era of digital civilization.

Data, algorithm, and scenes are the three core elements of governance technology. Block data value chain is to realize value integration beyond resource endowment, i.e., discovering and regenerating value with data flow through the hubs of block data, generating data drive, promoting and impacting technology flow, material flow, capital flow, talents flow, service flow, and optimizing resource allocation. Finally, a diversified value system including a whole business-based industrial chain, a whole society-based service chain, and a whole government-based governance chain are produced. Activation dataology, a data perspective and methodology of data storage and utilization reduced under the neuron scheduling system of block data, aims to offer a scheme for building a system of data fusion, algorithm, and scenes in this era of big data. Data search, correlative fusion, self-activation, hotspot reduction, and intelligent collision are interacting in the block data system and keeping cycling back and forth. With the magnification and reconstruction of data value, this process steadily promotes the spiral evolution of the whole system. As a theoretical hypothesis, activation dataology, like the "Eye in the Sky" toward deep big data universe, is a prediction of the arrival of cloud brain era for future humans, the revolution of big data thinking paradigm about chaotic data world beyond the "either/or", and "This and That" complex theories of determinism and probability theory, so it will predict uncertainty and unpredictability in a more accurate way. The arrival of cloud brain era driven by data, algorithm, and scenes will help us better understand the law of human social development.

5.2.2.2 Data Rights Law

From the first day when we came to know big data, we often regarded it as a new energy, new technology, and new way of organization, or as a new force shaping the future, creating more value with the cross-border, integration, openness, and sharing of data. However, open data and data flow usually lead to more risks. The excessive collection and abuse of personal information pose great challenges to the privacy of data subjects, information security of enterprises, social, or even national stability, thus arousing widespread and deep concerns about data sharing, privacy protection, and social justice, and becoming a major problem in global data governance. This puzzle led us to think deeper and try to come up with a theoretical hypothesis of "data persons". We call the rights derived from "data persons" as "data rights", the order on data rights as "data rights system", and the legal norms on data systems as "data rights law", so as to construct a theoretical framework of "data rights—data rights system—data rights law" (Key Laboratory of Big Data Strategy 2018).

Currently, there are no written laws on the protection of data rights in China, and the relevant provisions are stipulated in the Constitution, criminal law, civil law, and other laws (Table 5.2). *The General Provisions of the Civil Law*, adopted in March 2017, for the first time responds to the legal status of "data". Article 127 clearly stipulates such provisions shall prevail where laws provide for the protection of data and network virtual property. *"Technically, the clause is a cause clause or a referral clause. In general, however, there are specific provisions in relation to the arising clause"* (Weixing 2018). However, the Interpretation of *the General Provisions of the Civil Law of the People's Republic of China* compiled by the Legislative Affairs

Table 5.2 Relevant legal provisions on the protection of data rights in China

Category of law	Name of law	Implementation date	Relevant provisions
Constitution	Constitution	March 11, 2018	Article 33, Article 37, Article 38, Article 39, Article 41
Criminal law	Criminal Law	November 4, 2017	Article 253 Paragraph 1, Article 286 Paragraph 1, Article 287 Paragraph 1, Article 287 Paragraph 2
Civil law	Tort Liability Law of People's Republic of China	July 1, 2010	Article 2, Article 36
	Law of the PRC on Protection of the Rights and Interests of Consumers	March 15, 2014	Article 14, Article 29, Article 50
	General Provisions of the Civil Law of the People's Republic of China	October 1, 2017	Article 110, Article 111, Article 127
Other law	Passport Law of People's Republic China	January 1, 2007	Article 12, Article 20
	Law on Practicing Doctors of the People's Republic of China	August 27, 2009	Article 22, Article 37
	Statistics Law of the People's Republic of China	August 27, 2009	Article 9, Article 25, Article 37, Article 39
	Law of the People's Republic of China on Guarding State Secrets	October 1, 2010	Article 23, Article 24, Article 25, Article 26
	Law of the People's Republic of China on the Identity Card of Residents	January 1, 2012	Article 6, Article 13, Article 19, Article 20
	Law of the People's Republic of China on the Prevention and Treatment of Infectious Diseases	June 29, 2013	Article 12, Article 68, Article 69
	Postal Law of the People's Republic of China	April 24, 2015	Article 7, article 35, Article 36, Article 76
	National Security Law of the People's Republic of China	July 1, 2015	Article 51, Article 52, Article 53, Article 54
	Law of the People's Republic of China on Commercial Banks	October 1, 2015	Article 6, Article 29

(continued)

Table 5.2 (continued)

Category of law	Name of law	Implementation date	Relevant provisions
	Cybersecurity Law of the People's Republic of China	June 1, 2017	Article 10, Article 18, Article 21, Article 22, Article 27, Article 37, Article 40, Article 41, Article 42, Article 43, article 44, Article 45, Article 66, Article 76
	Law of the People's Republic of China on Lawyers	January 1, 2018	Article 38, Article 48
	Electronic Signature Law of the People's Republic of China	April 23, 2019	Article 15, Article 27, Article 34
	Code Law of the People's Republic of China	January 1, 2020	Article 1, Article 2, Article 7, Article 8, Article 12, Article 14, Article 17, Article 30, Article 31, Article 32

Committee of the NPC clearly states "In view of the complexity of data and network virtual property, confined by the chapter structure of the general provisions of the civil law, how to define data and network virtual property and how to specify data rights and network virtual property attributes and rights contents shall be specified by particular laws". In other words, *The General Principles of Civil Law* only raises the question of data rights, but does not make specific provisions. *The Civil Code of the People's Republic of China* (draft)[19] under compilation also covers data rights. Its "Personality Rights Compilation" has a special chapter with framework provisions on the right to privacy and the protection of personal information. However, from the perspective of legal adjustment objects, the protection objects are still "personal information" rather than "personal information rights". In this regard, Wang Limin, a famous civil jurist, suggested adding the word "rights" after "personal information" to clearly define "personal information rights", which, on the one hand, provide a legal basis for special laws, and on the other hand, fulfill the need for judicial protection of personal information.[20]

"All legal systems or legal theories can be divided into principles and technologies. The principles belong to fundamental values or the value foundation of the system, while the technologies are merely the means to realize the principles" (Bencai 2019). Based on the principles and theories of data rights law, we think that human rights, property rights, and data rights are three basic rights of human life in the future. Data

---

[19]Refer to *the Civil Code of the People's Republic of China (draft)* (December 16, 2019).

[20]Jin Hao. Wang Liming: The Right to Personal Information should be clearly stipulated in the Civil Code Personality Rights Draft. news.gm.com. 2019. https://news.gm.cn/2019-12/20/content_3341 8967.html.

rights are the synthesis of personality rights and property rights; the subjects of data rights are specific right holders, and the objects of data rights are specific datasets. The data rights break through the limitations of "one ownership for one object" and "Nothing can be an object of property, which has not a corporeal substance", and often manifests itself as multi-ownership of data. The data rights have the attributed quality of private rights, public rights, and sovereignty. The data rights system includes five basic dimensions: legal data rights system, data ownership system, public interest data rights system, usufruct data rights system, and sharing system. Sharing right is the essence of data rights; data rights law is a legal norm to adjust data ownership, data rights, data utilization, and data protection. Data rights law, reconstructing the new order of digital civilization, is a cardinal cornerstone of industrial civilization toward digital civilization, exploring and innovating the future rule of law, as well as enriching and deepening traditional civil law. Date rights law, with data confirmation as the core, not only use data to its full extent, but protect data rights as well. As a higher law, *Data Rights Law* is of advanced, scientific, and instructive significance to improve and practice the legal system of data rights composed of *Network Security Law*, *Data Security Law,* and *Personal Information Protection Law*. As the legal dimension of governance science and technology, data rights law is the necessity for the orderly circulation of data, the premise of data reuse, and the balance between personal privacy and data utilization in cyberspace. It is the basic material for constructing the "circumference" world of the legal empire in cyberspace. Data rights law and property rights law constitute two major legal foundations for the era of digital civilization.

5.2.2.3 Sovereignty Blockchain

"No cyber security, no national security" has become a broad consensus. In the era of big data, network security has been closely tied to national security and public security. For the vast majority of sovereignty states, another judgment is perfectly acceptable: "No cyber sovereignty, no cyber security". Network sovereignty and network security interpenetrate and crisscross each other. The establishment of international Internet governance rules via cyberlegislation must be guided by the concepts of national sovereignty and cybersecurity. A nation's construction, operation, and management of the Internet and its governance of crimes in cyberspace must be carried out under sovereignty and the rule of law. Blockchain as the fundamental technology of the bitcoin, on the one hand, has an "impossible triangle" in which the demands of "high efficiency and low energy", "decentralization", and "security" cannot be met simultaneously; on the other hand, encounters governance problems such as "switching center" rather than "decentralization" in nature, severe external dependence of data security technology, and difficult regulation of single technology (Key Laboratory of Big Data Strategy 2017). Technology has no position, but the people who control it have nationality. Once the blockchain governance agenda is set, rules made, and basic resource allocation rights controlled by technological powers, they will become the only masters of the blockchain world, and national relations will depend on technologies and reduce to the "law of the jungle" era of "the weak are the prey of the strong", which will be unacceptable and unbearable for modern civilized society.

Technology is not the law and cannot replace the law. The imagination expressed by governance technology is a kind of imagination that has borders, not "wild imagination" of super-sovereignty and no sovereignty. Respecting national sovereignty behind network sovereignty is a necessity for the development of blockchain. Different from the single technology governance of the blockchain, the sovereignty blockchain realizes the technology governance under the regulation of the law. The governance rules of sovereignty blockchain are generally composed of legal rules and technical rules. Legal rules are composed of legal framework, provisions, industrial policies, etc., which are authoritative under the rule of law. If they are violated, they shall bear corresponding legal responsibilities. Technical rules are composed of software, protocol, program, algorithm, supporting facilities, and other technical elements, which are essentially a series of machine-readable computer codes with rigid execution and irreversible characteristics. Only under the combination of the legal and technical rules and with consideration given to the authority of legal rules and the feasibility of technical rules can regulation and governance of sovereignty blockchain be more conducive to the interests of participants and the whole society, and the commercial application scenarios based on sovereignty blockchain technology, finally building a complete commercial system with the participation of regulators, commercial institutions, and consumers (China Blockchain Technology and Industrial Development Forum CBD-Forum 2016).

The shift from blockchain to sovereignty blockchain, on the one hand, provides new ideas, technologies, and models for governance in real society; on the other hand, extends the governance field to cyberspace; upgrades the joint governance of real society and network society; promotes social governance in a more delayering and interactive direction; and the functional, order, and institutional reconstruction of social governance. The invention of sovereignty blockchain shifts the blockchain from the rule of technology to rule of system, and upgrades from FinTech to governance technology. It is foreseeable that under the framework of governance technology, the innovative application of block data, data rights law, and sovereignty blockchain will play an unprecedented role and impact. In particular, the application of governance technology based on sovereignty blockchain in consultative democracy will offer new technical support and new path for the socialist political development with Chinese characteristics in a firm way, and trigger profound social reforms. The reforms mean scientific socialism has shown great vitality in China in twenty-first century, a great leap forward in China's socialist democratic political system, and the contribution of "Chinese wisdom" to human political civilization (Yuming 2018).

5.2.3 Cyberspace Governance Based on Sovereignty Blockchain

Cyberspace governance is an issue of the era. At present, the cyberspace is confronted with such problems as imperfect rules, irrational order, and unbalanced development. Meanwhile, it is also confronted with structural abnormalities, hegemony, and poor rule of law. In form, technology communities spontaneously make rules, but in fact, they are controlled, in the source, by the hegemony of the technology powers which form a monopoly of power under the "pseudo-decentralization" of the Internet. The Internet is not a place outside the law. The international community needs a just Internet rule of law system. The governance technology based on the sovereignty blockchain will innovate the governance model of cyberspace by virtue of its excellent characteristics such as governability, surveillability, decentralization, and multicenter in a comprehensive way. While maintaining cyberspace sovereignty, it will elevate the Internet of Value to be the Orderly Internet and build a Community with a Shared Future for Cyberspace.

5.2.3.1 Plight of the Internet Global Governance System

Internet governance is not only a major component of global governance, but also a key aspect of the game between great powers as well. At present, the global governance system of the Internet is in short supply, and the dilemma of "three unchanged tendencies" still exists in the rule of law. First, unchanged is the basic situation of cyberthreats, such as infringement of personal privacy, infringement of intellectual property rights, and encroachment of information resources. Second, unchanged is the basic situation of frequent network security incidents, e.g., network monitoring, network attacks, and cybercrimes. Third, unchanged is the basic pattern of global public hazards, such as cyberterrorism, cyberhegemony, and cybermilitarism (Zhenfeng 2016). The global governance system of the Internet is in a state of sickness, with poor institutions and in urgent need of updating and upgrading.

Global Internet Operation and Management Imbalance. Just like the real space, the operation of cyberspace needs to allocate and consume resources. IP address, domain name, port, protocol, and so on are the core resources necessary for Internet operation. These resources can neither be created out of thin air nor used at will, but need to be allocated and managed by specialized agencies. According to incomplete statistics, the main operation and management institutions of global Internet at present are ICANN (the Internet Corporation for Assigned Names and Numbers), RIRs (five Regional Internet Registration Management Institutions), ISOC (the Internet Society), IAB (Internet Architecture Board), IETF (Internet Engineering Task Force), IRTF (Internet Research Taskforce), ISO (International Organization for Standardization), the W3C (World Wide Web Consortium), INOG (Internet

Network Operators Group), etc.[21] Providing powerful technical support for global Internet, these institutions have the absolute right of management and domination, and control the essential infrastructure for global Internet, as well as the core standards and major agreements at management and technology. They have been a "dominant" "jungle society" under the global Internet governance system; however, they are basically in the hands of Western technological powers led by the United States, and their members are mainly the citizens of Western developed countries, which has formed a serious imbalance at the root. This imbalance, in turn, leads to the dilemma of the lack of discourse power of some stakeholders. It is mainly manifested in two things that are unguaranteed. First, unguaranteed are the lawful rights and interests of weak nations in technology and their citizens as the regulatory agencies and personnel are mainly composed of European and American nations and their citizens. Second, unguaranteed are the autonomy and voice of the Internet development path of weak technology nations as the global Internet governance system becomes monopolistic.

Network Hegemony and Domination. As the birthplace of the Internet, the United States owns and manages the most developed Internet technology and critical infrastructure worldwide, controls the production of major global information products and the management of Internet address resources and root servers, and has the absolute control right unmatched by other nations. Meanwhile, the control of cyberspace is almost controlled by the United States alone, and other nations, including China, are basically in the gray zone of cyberspace sovereignty, that is, in the chaotic state of "semi-sovereignty" or even "non-sovereignty" of cyberspace. In addition, the United States preaches "double standards"[22]: technological liberalism and dictatorship, giving leeway to terrorists and militarists. Although the weakening and restriction of national sovereignty by the new sci-tech revolution is the same for all nations, it is unbalanced and unequal for developed and developing countries with different technological levels (Xudong 1997). Such imbalance and inequality, the dominated and being dominated international relations under cyberhegemony, are extremely detrimental to international justice, especially to the third world nations.

Cybersecurity and Cybercrimes. Cybersecurity and cybercrimes are other hard nuts in global Internet governance. The Internet is virtual, anonymous, cross-border and borderless, open and decentralized, and instantly interactive in nature. These characteristics set a possible "hotbed" for criminals to carry out anonymous network attacks, fraud, pyramid selling, and other illegal and criminal activities. There are two forms of cybercrime: the first is to monitor networks, attack websites, spread viruses, and other illegal intrusion and destruction, such as "Snowden Affair", "Five Eyes Alliance Incident", and "Stuxnet Incident". The second is the traditional crimes

---

[21] To a certain extent, institutions and organizations involved in the global Internet are APEC, International Organization of Southeast Asian Nations, Council of Europe, European Union, Forum of Incident Response and Security Teams, Group of Eight, Institute of Electrical and Electronic Engineers, International Telecommunication Union, Internet Governance Forum, INTERPOL, Meridian Process, North Atlantic Treaty Organization, Organization of American States, Organization for Economic Cooperation and Development, etc.

[22] "Double standards" refers to the practice of a double standard on cyberfreedom and cybersecurity, that is, one set of standards for oneself and Allies the other set for developing countries.

carried out via the Internet, such as Internet financial fraud, illegal fund raising, network theft, false advertising, cyber manhunt, insults and slanders, and online espionage, challenging existing global security system. Compared with the traditional crimes, the network crimes are featured by the following aspects. First, strong destructive power and the growing number of adolescents committing crimes. Second, low cost of crime, more victims, and serious economic losses. Third, widespread danger involving all walks of life and all fields. Some criminal activities even endanger national security in politics, economy, and society.

5.2.3.2 International Frictions Over Cyberspace Sovereignty

Cyberspace is undoubtedly an important field in the exercise of cybersovereignty. However, "due to the limited understanding and relevant practices of nations in cyberspace, more seriously, due to the divergences and even oppositions in ideology, values and practical national interests", (Zhixiong and Yahui 2017) the international community still faces differences in many areas of cyberspace. On the whole, the international community is divided on cognition, strategy, and governance in cyberspace.

Cognitive Divergence: Global Commons and Sovereign Territories. "Global Commons" theory, represented by the United States, Japan, the European Union, and other cyber developed countries and organizations, holds that cyberspace, unlike physical space, is not under the jurisdiction or domination of any single nation and shall be regarded as international Commons such as high seas and space. Take the United States for example. It compares cyberspace with high seas, international airspace, and space, and classifies cyberspace into the "Global Commons"[23] that cannot be reached by a single sovereignty state. It believes that the management of cyberspace shall transcend the boundary between sovereignty states in the traditional sense, and nations shall not exercise sovereignty in cyberspace. Different from "Global Commons" theory, the "Sovereignty Territory" theory, represented by Russia, Brazil, the Shanghai Cooperation Organization, the Asia–Pacific Security Cooperation Council, and other emerging cyber nations and organizations, holds that cyberspace has sovereignty attribute and nations shall establish and exercise their sovereignty in cyberspace. For example, Russia, together with China and other SCO member states, submitted *the International Code of Conduct on Information Security* to 66th General Assembly in 2011, believing that the right to make decisions on public policy issues related to the Internet is the sovereignty of all nations, and that the right to international discourse and governance in cyberspace shall be respected. The confrontation between the two propositions in the current international community is becoming more intense. More and more nations, inclined to the "Sovereignty Territory" theory, argue that national sovereignty needs to be exercised in cyberspace. However, nations have obvious cognitive divergences on the

---

[23]Global Commons is the domain or resource on which the security and prosperity of all nations depend, not at the mercy of any one nation.

attributes of cyberspace in practice due to the significant gap in development, history, culture, and social systems (Ying and Ling 2019). Consequently, nations still have divergences on cyberspace sovereignty in the formulation of international rules of cyberspace.

Governance Differences: Multistakeholder Governance and Multilateralism Governance. At present, cyberspace governance presents two camps: developed countries led by the United States and developing countries represented by China, Russia, and Brazil. The former supports multistakeholder governance, while the latter multilateralist governance, while "multistakeholder" is the "recognized" governance model of global cyberspace governance currently. Its supporters advocate that "technology experts, commercial institutions and civil society shall take the lead in cyberspace governance, and governments shall not interfere too much, and even inter-state governmental organizations such as the United Nations shall be excluded" (Mingjin 2016). This model holds that "the decentralized global cyberspace communication has deprived the government of its central dominant position in traditional governance theories", (Wenming 2017) and advocates that Internet governance shall be "bottom-up". Superficially, multistakeholder governance model plays a certain role in taking into account the interests of all parties, but this model is difficult to achieve effective governance of cyberspace due to the lack of cooperation and support from sovereignty states. Different from the multistakeholder governance model advocated by cyber developed countries, cyber developing countries tend to be government-led and advocate strengthening cyberspace governance through the United Nations or other international organizations. Known as the "multilateralism governance model", this model advocates national "top-down" governance in cyberspace and that "The principle of state sovereignty in cyberspace and the solution of disorderly problems in cyberspace should be centered on nations. Nations have the right to safeguarding digital sovereignty and national security in cyberspace. Some entity organizations with the nation as the subject of governance shall be established within the framework of the United Nations to coordinate cyber governance issues" (Wenming 2017). In a nutshell, the divergence between the multistakeholder governance model and the multilateralist governance model is a dispute between the defenders and reformists of the governance mechanism of cyberspace based on their respective interests. It is foreseeable that divergences and games around this issue will last for a long time" (Kun and Qichao 2019).

Strategic Divergence: Cyberliberalism and a Community with a Shared Future for Cyberspace. For the sake of values, Western countries advocate "human rights above sovereignty", liberalism in cyberspace network, and the idea that basic human rights are inviolable. Against the extension of real space control to cyberspace control, they believe network borders bring challenges to democracy, and they do not accept any activity impeding the free flow of information. For example, the United States believes that states cannot for any reason impede the free flow of connectivity and data, and that fundamental freedoms in cyberspace shall be guaranteed. To this end, the United States issued *The International Strategy of Cyberspace* and *The Action Strategy of Cyberspace* in 2011 to promote cyberspace freedom. These two documents constituted the overall framework of the international strategic system of the Internet of the United States, within the framework of cyberliberalism theory.[24] Contrary to the argument of cyberliberals, cyber developing countries believe that "cyberliberalism does not meet the needs of cyberspace", (Wang Mingjin 2016) the norms governing international relations and *the UN Charter* in cyberspace shall be taken as the fundamental basis, the territorial integrity, political independence and human rights freedom of all nations be respected, the principle of national security and sovereign independence be followed, and no nation can promote cyberhegemony under the banner of "Internet freedom". China, for example, attaches great importance to cyberspace sovereignty. In recent years, China has actively advocated respecting and safeguarding the sovereignty of other nations in cyberspace, and taken this as one of its core propositions on international law and order in cyberspace. At the Second World Internet Conference in December 2015, President Xi Jinping first proposed the concept of "building a Community with a Shared Future for Cyberspace" and elaborated on the "Four Principles" and "Five Propositions" for building a Community with a Shared Future for Cyberspace. A Community with a Shared Future for Cyberspace serves the interests of most nations and has been accepted by a growing number of nations worldwide since it was proposed. However, some Western countries still cast doubts over this idea, out of ideology.

5.2.3.3   A Community with a Shared Future for Cyberspace

The highest voice of the new era reverberates throughout China and the international community. "A Community with a Shared Future for Cyberspace is a Chinese approach for global cooperation and governance in cyberspace, and it utters China's voice for the prosperity and security of global cyber culture" (Feng 2018). In retrospect of three industrial revolutions, there are two noticeable characteristics. First, new technological means promoted a series of new inventions, raising human productivity and the scope of human activities. Second, with new technological means, mankind has tapped more into potential and innovated the whole society. The third Industrial Revolution, marked by Internet technology, has turned the world into a global village where people separated thousands of miles are no longer "invisible". Information technology, represented by digital twins, has ushered in new changes in social production, new space in human life, new areas in national governance, and human understanding and ability to understand and transform the world. China's view on global governance advocates democracy in international relations and the

---

[24] Among them, the International Cyberspace Strategy takes the freedom of cyberspace as a core concept and a major component, arguing that "The international cyberspace policy of the United States reflects the fundamental principles of the United States, namely the core commitment to fundamental freedoms, personal privacy and the free flow of data". "After network liberalism theory, it has become the official ideology of the US government and is regarded as an indisputable so-called "universal value". With the strong communication of the Western discourse system represented by the US, it basically dominates the research and discussion on Internet issues in the years to come. (Li Chuanjun, Order Change and Model Construction of Global Governance in Cyberspace, *Journal of Wuhan University of Science and Technology (Social Science Edition)*, 2019, No. 1, pp. 20–25.).

equality of all nations, big or small, strong or weak, rich or poor, which must be based on and premised on a Community with a Shared Future for Cyberspace. Only with a Community with a Shared Future for Cyberspace can we address such issues as unbalanced Internet development, irrational rules and order, encourage the United Nations to play a positive role, give developing countries greater representation and say in international affairs, and facilitate their participation on an equal footing in the reform and development of global governance system.

The concept of a Community with a Shared Future for Cyberspace dates back to 2015. With rich connotation and far-reaching impact, it is an essential component of President Xi Jinping's new concept, new thought, and new strategy of cyberspace governance (Table 5.3), a scientific path to ensure national data security and comprehensive governance of cyberspace in the new era. From the theoretical perspective, it is a network governance strategy of collaborative cooperation and shared responsibility put forward in the Internet era when humans are faced with network risks beyond geographical boundaries (Hui and Jiali 2017). It inherits, deepens, and develops the Open Theory of Marxism and the Theory of World Communication. From the perspective of value, its vision conforms to the overall layout of the five-sphere integrated system, so it is of great value to China and the world in terms of economy, politics, culture, society, and ecology. It is conducive to economic development and common prosperity of all nations, the enhancement of China's international discourse power, cultural exchanges, and mutual learning between nations, maintaining global stability and purifying network ecology. From a practical perspective, it serves the common interests of people worldwide to build a fair and reasonable global Internet governance system, integrate the rule of law with the rule of virtue, safeguard the common interests of people worldwide, and build a Community with a Shared Future for Cyberspace with joint efforts (Jianmei 2019).

"Symbiosis is the basis for operating a Community with a Shared Future for Cyberspace. Common security is its operating environment, equal autonomy its model, multi-dimensional cooperation its mechanism and interests sharing its goal" (Suibing 2018). The cyberspace governance scheme based on the sovereignty blockchain contributes to an Internet governance system featuring equality, consensus, and co-governance, and provide an operating environment and technologies for a Community with a Shared Future for Cyberspace. First, the sovereignty blockchain promotes equal cooperation in a Community with a Shared Future for Cyberspace. In a point-to-point cyberspace, it encourages nations to respect cybersovereignty,[25] their right to independently choose their own path of cyberdevelopment, model of cybermanagement, Internet public policies and partake in international cybergovernance on an equal footing; erase the gap between big and

---

[25]The significance of the transition from blockchain to sovereignty blockchain lies not only in the development of blockchain, but also in the new meaning of "sovereign governance" brought to cyberspace governance. Among the Four Principles for reforming global Internet governance system, the first principle is "respecting cyber sovereignty". The theory of cybersovereignty is the theoretical basis for the establishment of a Community with a Shared Future for Cyberspace and the logical starting point of the other three principles.

Table 5.3 President Xi Jinping's new concept, new thought, and new strategy of cyberspace governance at world internet conference

Year	Conference	Important propositions
2014	First World Internet Conference	"China is ready to work with other nations to deepen international cooperation, respect cyber sovereignty, safeguard cyber security, jointly build a peaceful, secure, open and cooperative cyber space and establish a multilateral, democratic and transparent international Internet governance system based on the principle of mutual respect and trust"
2015	Second World Internet Conference	"Cyberspace is the common space of human activities, and the future of cyberspace should be in the hands of all nations. Nations should strengthen communication, expand consensus, deepen cooperation and jointly build a Community with a Shared Future for Cyberspace"
2016	Third World Internet Conference	"The Internet is the most dynamic area of our era. Rapidly developing Internet has brought profound changes to human production and life as well as new opportunities and challenges to human society. The Development of the Internet is nationally and geographically borderless. To make good use of, develop and govern the Internet, we must deepen international cooperation in cyberspace and jointly build a Community of a Shared Future for Cyberspace". "China is ready to work with the international community to uphold the common well-being of mankind and the concept of cyber sovereignty, make global Internet governance more just and equitable, and achieve the goals of equal respect, innovative development, openness and sharing, security and order in cyberspace"
2017	Fourth World Internet Conference	"A new round of scientific and industrial revolution currently, represented by information technology, is emerging, injecting strong impetus into economic and social development. Meanwhile, the development of the Internet has brought many new challenges to the sovereignty, security and development interests of nations worldwide. The reform of the global Internet governance system has entered a critical period, and the initiative of a Community of a Shared Future for Cyberspace has increasingly become the broad consensus of the international community". "Advocating the Four Principles and the Five Propositions is to work with the international community to respect cyber sovereignty and inherit the spirit of partnership. We shall discuss our issues and work together for common development, security, participation in governance and achievement sharing"
2018	Fifth World Internet Conference	"The world today is undergoing a broader and deeper technological revolution and industrial change. Breakthroughs have been made in modern information technologies such as the Internet, big data and artificial intelligence. Digital economy is booming and the interests of all nations are more closely linked. It is imperative for us to accelerate digital economy and make the global Internet governance system more just and equitable". "Nations worldwide may have different national conditions, different stages of Internet development and different practical challenges, but they share the same desire to develop digital economy, the same interest in addressing cyber security challenges and the same need to strengthen cyberspace governance. Nations should deepen practical cooperation, take common progress as the driving force and win–win results as the goal, blaze a path of mutual trust and governance, and make the Community of Shared Future for cyberspace more vibrant"

(continued)

Table 5.3 (continued)

Year	Conference	Important propositions
2019	Sixth World Internet Conference	A new round of sci-tech revolution and industrial transformation currently is accelerating, and new technologies, applications and business forms, e.g., artificial intelligence, big data and the Internet of Things are on the rise. The Internet has gained stronger momentum and broader space for development. It is the common responsibility of the international community to develop, utilize and govern the Internet to bring more benefits to mankind. "All nations should follow the trend of the times, shoulder the responsibility for development, jointly meet risks and challenges, advance global governance in cyberspace and build a Community with a Shared Future for Cyberspace"

small nations, strong and weak nations; and build an open and transparent Community with a Shared Future for Cyberspace with high degree of information security. Second, the sovereignty blockchain establishes a new consensus system for a Community with a Shared Future for Cyberspace. The use of sovereignty blockchain realizes data exchange and information sharing among nations; upgrade their rights to know, participate, express, and take the initiative in consultation and discussion; help build impersonal trust and consensus established by codes, protocols, and rules; and break down Internet trust barriers among nations. Finally, the sovereignty blockchain promotes a Community with a Shared Future for Cyberspace. The sovereignty blockchain helps build a new model of Internet governance, encourage nations to work together for Internet governance and the governance capacity of a Community with a Shared Future for Cyberspace, as well as a sound order in cyberspace.

5.3   Tech for Social Good and Rule of Conscience

Only by striking a balance between "good and evil, justice and benefit" can we achieve sustainable development. The power of science and technology itself is enormous, and its development is increasingly rapid. How to make good use of it will impact the well-being of human society. Science and technology are the expression of human nature, the means of harmony between humans and nature, the integration of goodness and conscience in human nature, and the objective truth of the outside world. The "Tech for Social Good" we refer to a kind of "people-oriented" choice of conscience oriented toward beauty and light. Choosing "Tech for Social Good" means improving sci-tech capabilities steadily, providing better "good" products and services for humans, and improving production efficiency and life quality, and doing what is necessary.

5.3.1 The Value Orientation of "Data Persons"

What is a person? It has always been the fundamental problem for humans to reflect on and study themselves. "Human is a unique being who tries to recognize his uniqueness. He is not trying to recognize his animal nature, but his human nature. He doesn't seek his own origin, but his own destiny. The gulf between human and non-human can only be understood from the perspective of humans" (Herschel 1994). The proposal of "data persons" not only transcends the traditional hypothesis of human nature, but redefines sci-tech ethics and "Tech for Social Good". Studies on the values of "data persons" are of great significance to the value selection of natural persons, the value orientation of gene-edited men, and the value design of robots in the sixth round of Kondratieff Wave.

The Assumption of "Data Persons". The assumption of human nature is a process of selectively abstracting the expression of the defined body of human nature based on certain value orientation. Generally speaking, the assumption of human nature, as a presupposition, is used to derive and deduce a certain theoretical system. The assumption of human nature should serve the theoretical system, which has a certain value orientation. When the constructor chooses the assumption of human nature, he must identify with the same value orientation, so as to make a certain theoretical system consistent. This value orientation is contained in the whole process of the theoretical system and finally embodied in practice. In the era of big data, everything is "online" and everything can be quantified. All humans, machines, and objects shall exist, be connected and jointly create value as "data persons". Individuals will leave "data footprints" in various data systems. A person's characteristics can be restored to form a "data persons" through correlation analysis. From economic persons, social persons to data persons, human nature assumption has different types in different times, reflected in various stages and different patterns of human nature assumptions. When he makes the assumption of human nature, the historical background is the decisive factor that the constructor must take into consideration. Therefore, epochal character is a very important feature of human nature assumption. "Data person" is not only the datalization of humans, all objects and parts will also exist as data individuals and interact with each other. "We are at a time when technological progress brings about changes in human nature" (Fang 2013). On the one hand, the social relationship changes brought by artificial intelligence, 3D printing, gene editing, and other emerging technologies will make humans surpass the limits of evolution. On the other hand, humans have to face the situation that "by the middle of this century, non-biological intelligence will be one billion times the intelligence of all people today" (Handong 2017). At that time, robots and gene-edited men will exist as "data persons", coexisting, interacting, and complementing with natural persons. They will become the "three main bodies" of human society in the future. The "data persons", as a new assumption of humanity, exists in the context of data civilization. Selecting, guiding, and designing the value orientation of natural persons, robots, and gene-edited men with the concept of data civilization and "data persons" transcend the traditional boundary of good and evil, and free traditional restriction on organization effects. It can be said that compared with the assumption of human nature such as economic men, social men, and complex men, the assumption of "data persons" can better adapt to the theoretical and practical requirements of data civilization, and better realize the liberation of all mankind and the free and all-round development of humans.

Values of "Data Persons": Altruism. The idea of egoism put forward by Adam Smith in *The Wealth of Nations* and the idea of altruism[26] in *The Theory of Moral Sentiments* constitute the classic "Smith Paradox". There is no doubt that human nature includes inherent egoism and altruism. Altruism, a manifestation of human virtue, is a theory of ethics, generally referring to the life attitude and behavior principle that put social interests first and sacrifice personal interests for social benefits. As early as nineteenth century, Comte, a French positivist philosopher, proposed the concept of "altruism" and used it to explain the selfless behavior in society. Comte believed that "'altruism' is a desire or trend to live for others and a tendency opposite to self-interest" (Auguste 1875). "Just as people have rational demands on their thoughts, so do they on their actions, and altruism is one of the rational demands on behavior" (Nagel 1978). Therefore, altruism first emphasizes the interests of others and advocates the dedication to others' welfare at the expense of self-interests. At present, altruism is generally considered to be characteristic of voluntarily helping others without seeking future rewards. The interdisciplinary study of game theory and evolution theory shows altruistic groups have more evolutionary advantages in ecological competition than selfish groups. "Data persons" emphasize altruism in the way people behave and exist. The function of "data persons" is to help humans create a shareable and public big data field. Its instrumental value determines the natural altruistic characteristic of "data persons". If the altruism of "data persons" can bring more benefits and convenience to people, then people will exhibit more altruistic behaviors based on the pursuit of interests. Darwin wrote in *The Descent of Man*, "A tribe with many of its members willing to help each other and sacrifice themselves for the good of the group, will prevail over other tribes" (Darwin 1997). The altruistic nature of "data persons" also helps promote cooperation between humans, as, based on the altruistic nature of "data persons", even though only a small number of people benefit from the cooperation at first, but with the participation of more and more people, altruistic behaviors have risen from accidental cooperative relations to specific legal relations, promoting humans to obtain continuous benefits from altruistic behaviors. To this end, a nation, based on its aspiration to promote social well-being and human progress, needs to create protective mechanisms for "altruistic behavior", that is, altruistic values. Under the impact of this value, people will form an ideological consciousness that produces beneficial effects on others materially or spiritually through behaviors and activities, and finally build a more harmonious society.

---

[26]In *The Theory of Moral Sentiments*, Adam Smith made clear in Chapter 1 "Of Sympathy", "How selfish soever man may be supposed, there are evidently some principles in his nature, which interest him in the fortune of others, and render their happiness necessary to him, though he derives nothing from it except the pleasure of seeing". ([UK] Adam Smith. *The Theory of Moral Sentiments*, translated by Jiang Ziqiang et al., The Commercial Press, 2015, p. 5.).

Fig. 5.1 Relationship
between altruism and sharing
behavior

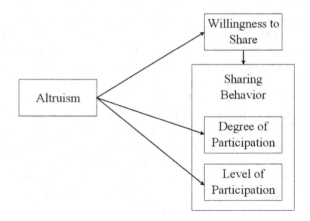

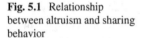

Fig. 5.1 Relationship between altruism and sharing behavior

Sharing: Data Culture of Altruism. From agricultural civilization to industrial civilization and then to digital civilization, human society has been steadily progressing with science and technology, and human production activities and lifestyle have been more shared. Especially with the rise of Open Access[27] Movement and sharing economy in recent years, as a new development concept, sharing has expanded from science and technology to economy, society, ideology, culture, etc., and people come to see clearly the importance of sharing to the common life and development of mankind (Key Laboratory of Big Data Strategy 2018). Altruism, with the behavioral tendency and value proposition of benefiting others, is an externalized conscious practice aimed at promoting individual willingness to share and their sharing behavior (Fig. 5.1). Sharing behavior includes the level and degree of participation, which can directly reflect individual differences in sharing behaviors. Every individual is both a beneficiary and benefactor to the community and the public. Through sharing and mutual benefit, communities will become more harmonious and sustainable. Jack Ma shared his idea at a conference that the world needs to make good use of DT (data technology) as its core of is altruism, "To believe others are more important, smarter and more capable than you, and to believe only when others succeed can you succeed". Alibaba Group is not an e-commerce company, but a

---

[27]Open Access Movement, a knowledge-sharing model and a scientific movement launched at the turn of 1990s and 2000s, was aimed at promoting the sharing of scientific research results. Among them, the Budapest Open Access Initiative issued in 2001, the Bethesda Declaration on Open Access signed in 2003, and the Berlin Declaration on the Open Use of Natural science and Human Science Resources are the embodiment of the concept of shared development. According to the Budapest Open Access Initiative, open access refers to the fact that scientists upload their research literatures on the Internet, providing free access to anyone to read, download, copy, print, publish, and retrieve. Or they will set up link and index to transform them into software, data, or any other legal forms, so as to make people freely access to and use the research from the Internet. On the one hand, Open Access Movement made scientific data freely available to the public, breaking through the price barrier of knowledge; on the other hand, it expanded the availability of scientific research results and overcame the barriers of access to scientific documents. (Hu Bo. The Sharing Model and the Future Development of Intellectual Property—Review on the Alternative Model of Intellectual Property. *Law and Social Development*, No.4, 2013, pp. 99–111.).

company that helps others operate e-commerce. If you want to be successful, you must be altruistic first. It is only after altruism that one can benefit oneself. The most typical example of sharing culture is the Internet. "The Internet holds the spirit of sharing since its birth, manifested by information sharing, technology sharing, and distribution on demand" (Di 2014). "Equality and sharing are the souls of the Internet. The 'body' of the Internet, technical framework, communication protocol and terminal equipment is being constantly updated while the 'soul' is consistent, up to now" (Ning and Shujun 2018). Sharing culture, as a mainstream culture, affects the entire society in the era of big data and implants endless power and energy into social development. Adam Smith stated in *The Theory of Moral Sentiments* that "If a society of the fruits of economic development cannot really divert to the hands of the public, it is unpopular in a moral and a risk. because it is bound to pose a threat to social stability" (Smith 2008). If culture doesn't stop loss, the effect of an economic stop loss is limited. The data culture of altruism has finally removed the bloodstained primal urge for capital accumulation, allowing the fruits of economic development to be distributed to the public through data-driven organization and sharing. This is not only reflected in the public innovation and entrepreneurship reflecting collective wisdom it advocates, but also in the pleasure and comfort enjoyed by many "unicorn companies" in a transparent data environment (Key Laboratory of Big Data Strategy 2016).

5.3.2 The Soul of Science

Sci-tech development is an important symbol of a nation's comprehensive national strength and international status. The power of science and technology has become the nation's most important strategic resource and a powerful driving force of social development. In the face of the sci-tech development in the next 80 years, three basic judgments shall arouse our thinking and attention. First, the integration of natural science and philosophy and social science is an inevitable trend in the second half of twenty-first century. Second, it's an inevitable choice to guide the development of natural science by philosophy and social science. Third, it's an inevitable requirement of the digital civilization era to accompany the truth of science with the goodness of humanity.

5.3.2.1 The Integrated Development of Philosophy and Science

"Philosophy and social sciences are an important tool for people to understand and transform the world, and an important force for historical development and social progress. Its development reflects a nation's thinking ability, spiritual character and civilized quality, a nation's comprehensive strength and international competitiveness. How a nation develops depends on natural science and philosophy and social science. A nation without a well-developed natural science can't be at the forefront of the world, nor can a nation without a prosperous philosophy and social science be at the forefront of the world" (Jinping 2016). The relationship between philosophy and social science and natural science is, in the final analysis, the relationship between human and objects, spirit and material. One is to build a home for spiritual life, the other to build a home for material life. The two are interdependent, interlinked, and mutually reinforcing.

Mr. Cai Yuanpei, a famous Chinese educationist and scientist, once discussed the interrelationship between science and philosophy in this way, "If we abandon science and study philosophy, it tends to be groundless assumption. If we stay away from philosophy and study science, we will inevitably confine ourselves. The two can be distinguished but not separated. The most moderate statement is to take philosophy as a universal science, and to seek the universal law by synthesizing all the laws in sciences and removing contradictory points between them. The application of universal knowledge to every science is a method or premise in exploring the highest ontology and checking it" (Cai Yuanpei Research Association of China 1997). Philosophy's pursuit of universal laws must be based on natural sciences. And as a universal knowledge, philosophy must be helpful to natural science from the fundamental aspects in terms of methods and premises (Xiaoli 2003). Therefore, we can neither research philosophy without science, nor solely research without philosophy.

Global sci-tech innovation has been more intense and vigorous since the beginning of twenty-first century. A new round of sci-tech revolution and industrial transformation is restructuring the global landscape of innovation and reshaping global economic structure. The new generation of information technology represented by artificial intelligence, quantum information, mobile communication, Internet of Things, and blockchain accelerates breakthroughs. The life sciences represented by synthetic biology, gene editing, brain science, regenerative medicine, etc. are gestating new changes. Advanced manufacturing technologies that integrate robotics, digitization, and new materials are accelerating the transformation of the manufacturing industry into an intelligent, service-oriented, and environment-friendly one (Jinping 2018). There is a growing trend toward cross-merging between disciplines, between science and technology, between technologies, and between natural and philosophy and social science.

Cross-border cooperation, the foundation for the era of big data, requires people to recognize the whole society as a system. On the basis of highly developed natural science and philosophy and social science, a high degree of integration has appeared. "Science and philosophy are originally of the same source" (Hua 2019). "Knowledge is like a tree, with philosophy as the root, science as the branch" (Descartes). "The relation between philosophy and science is the relation between universality and particularity. Philosophy is universality and science is particularity" (Stalin). If the integrated development of natural science and philosophy and social science in the era of big data is only a "small trial", then in the second half of twenty-first century, it will be given "full play". "In the universe and the world, human is the 'biggest variable' and object is the 'biggest constant'" (Qingqing 2016). Philosophy shapes the spirit and science discerns the matter. Philosophy is an important instrument for humans to "understand the world" and "transform the world". The integration of philosophy and science will continue to write wonderful chapters with titles such as "How far we can think determines how far we can go" and "How strong we are determines how high we can fly".

5.3.2.2 Philosophy Guides the Flourishing of Science

Philosophy and social science, a theoretical system to explore and summarize the law of human society, reflect the cultural-ethical conditions and civilization quality of a nation (Guanyi 2013). A great era is bound to bring about the prosperity of philosophy and social science. The prosperity of a great nation is also inseparable from the prosperity of philosophy and social science. It can be said that every major leap forward in human society and every major development of human civilization are inseparable from the intellectual transformation and ideological guidance of philosophy and social science. Philosophy and social science play an irreplaceable important role in the development of natural science from beginning to end.[28]

First, the philosophical worldview plays a pivotal role in dominating scientific activities and philosophy has a guiding role in natural science, which is an important principle of Marxism. Engels called this "guiding" role of philosophy in natural science as a "dominant" role.[29] He stated "No matter what attitude natural scientists adopt, they are still subject to philosophy. The only question is whether they are willing to be governed by some bad and fashionable philosophy, or by a theoretical thought based on the history and achievements of knowledgeable mind" (Marx and Engels 1972). As any scientific research involves the relationship between humans and nature, that is, the relationship between scientific workers and objects being recognized and transformed, as well as with social environment and social conditions. How to deal with the relationship between subject and object is involved with guiding philosophical ideology.[30] In addition, science is a cognitive activity aimed at discovering, studying, and understanding the nature and laws of objective entities. But no objective entity exists in isolation, and it is always intricately interwoven with other phenomena, secondary factors and accidental factors, with its essence deeply hidden behind things, which requires the subject of cognition to separate the essence from the intricate phenomena with dialectical thinking.

---

[28]Fu Zhenghua, Philosophy: Yeast in the Development of Science and Technology—On the Influence of Philosophy on the Development of Science and Technology. *Journal of Jingmen Vocational Technical College*, 1999, No. 5, pp. 61–66.

[29]Xiong Keshan, Zhao Shuangdong. The Need to Adhere to the Guidance of Marxist Philosophy to the Natural Sciences. *Journal of Qingdao Agricultural University (Social Science)*, 1987, No. 1, p.1.

[30]Therefore, Lenin called on that "Scientists of natural sciences should be modern materialists and self-conscious supporters of materialism represented by Marx, i.e. dialectical materialists". ([Russia] Lenin. *Selected Works of Lenin* (Vol. 4), translated by The Compilation and Translation Bureau of the Works of Marx, Engels, Lenin, Stalin. People's Publishing House, 1972, pp. 609–670.).

Second, philosophical methodology plays a major guiding role in science. Methodology and worldview are unified. Methodology determines worldview and the worldview in turn determines methodology. The guiding role of philosophical methodology in science is manifested as the dominant role of the philosophical world-view in scientific activities. The difference is that methodology is more concrete in guiding science. The three laws of materialist dialectics[31] can be applied to the study of natural science. They are the most basic methodological principles of natural science methodology.[32] In addition, the guiding role of philosophical methodology in science is also embodied in some specific thinking forms of dialectical logic, such as induction and deduction, analysis and synthesis, hypothesis and proof, and unification of history and logic. There is no doubt that philosophy serves a guiding role in scientific research, in collecting or collating empirical material, in proposing theoretical hypotheses or constructing theoretical systems.

Third, the critical spirit and skeptical spirit of philosophy are the solid ideological foundation of scientific innovation. Philosophy, critical in nature, with a restless soul never satisfied with reality, not only describes the phenomenon, but also uses a crit-ical perspective to examine and evaluate the relationship between people and the real world, based on which it then guides people to transform this reality relationship.[33] Similarly, philosophy is also skeptical in nature, for skepticism is the inherent quality of philosophy. Marx has a famous saying, "to doubt everything". The spirit of skep-ticism is essentially consistent with the critical spirit of philosophy. Only by asking why in every event can we trace the roots and find the essence of things. Philosophy is characteristic of exploring the inside of the world, surpassing the reality and pursuing the infinity, which is inseparable from its skeptical spirit.[34] In essence, the critical

---

[31] There are three laws of materialist dialectics: the law of unity and opposites, the law of quantitative change and qualitative change, and the law of negation of negation. The universality of these three laws in philosophy has reached their limits. Hegel first expounded this in *Logic*, and Engels summarized and refined it in *Logic*, thus making the law of dialectics clearer.

[32] That is to say, philosophical methodology is embodied prominently in the three laws of materialist dialectics. "For the law of dialectics is abstracted from nature and the history of human society, the law of dialectics is nothing but the most general law of these two aspects of historical development and of thought". Since the laws of dialectics are abstracted from the history of the development of the objective world, including nature, the laws of dialectics are "also valid for the theoretical natural sciences".

[33] Marx once had a brilliant exposition: "In the positive understanding of existing things, dialectics also contains the negative understanding of existing things, that is, the understanding of the inevitable demise of existing things; Dialectic understands each established form in its transitory aspect as well as in its constant motion; Dialectics doesn't worship anything; by its very nature it is critical and revolutionary". At the beginning of his theory, the founder of Marxism declared that he would mercilessly criticize everything existing and find a new world in the criticism of the old world. Only criticism can break the shackles of habitual force and thinking pattern; only criticism dares to challenge the authority of theory; only criticism can emancipate the mind constantly, so as to discover, invent, and create something, and constantly push science forward.

[34] Einstein once said that it was Mach's indestructible skepticism that led him to explore time and space. Mach said "Others figure out what space and time are at a very young age; I have not yet figured it out when I am grown up, so I have been trying to figure it out. As a result, I go deeper than others".

spirit and the skeptical spirit of philosophy are a kind of creative spirit and creative consciousness. In other words, the critical spirit and the skeptical spirit are the inner requirements of creative thinking. Philosophical critical spirit and skeptical spirit are huge driving forces of science, which paves the way for science. Meanwhile, they also provide a solid ideological foundation for science.

5.3.2.3 Protecting the Truth of Science by the Goodness of Humanity

Science and technology are the primary productivity and revolutionary forces for the progress of human society. Every major breakthrough in science and technology brings about profound changes in economy and great progress in human society. Nowadays, science and technology penetrate all aspects of people's production and life, changing people's production methods and lifestyles, and even thinking and behavior in a thorough way. Never has human society been so deeply influenced and dependent on technology as it is today. Technology is trying to control the future and destiny of mankind, and a technological society is approaching the world with an irresistible trend (Qi 2019). Science and technology can do both good and evil. In its promotion of social progress, they also bring a series of unprecedented global issues such as "nuclear war", "cyber war", "financial war", "biological war", and "threat from non-sovereignty powers" to human existence and development.

"Science and humanity are woven from the same loom".[35] "Modern natural science is the daughter of humanism" (Windelband 1993). The negative effects of science are bound to be criticized by humanists who call for guiding the development of science with humanity and protecting the truth of science by the goodness of humanity (Wenfu 2016). The spirit of science is seeking truth, while the spirit of humanity is seeking goodness. The pursuit of truth by itself doesn't guarantee a correct direction. "In their hearts they thought they are doing good things, but they don't understand the true meaning of it when they have done them".[36] The development of science and technology, especially that of artificial intelligence and gene editing technologies, has brought about major problems such as "personal information leakage", "gene-edited babies", and "artificial intelligence evil". Some people even use sci-tech achievements to engage in illegal activities. For example, the

---

[35] It has been said that human quest for creativity requires two wings: science and humanity. Science explains every possible thing in the universe, so that we can understand more about the hardware in the universe; humanity explains everything that can be thought of in human thought. Humanity builds our software. Science tells us what it takes to achieve a goal chosen by man, and humanity can tell us where the future of mankind can go with the fruits of science.

[36] From *The Author's Preface of Tai Shi Gong* by Sima Qian. The context reads: "为人臣子而不通于《春秋》之义者, 必陷篡弑之诛、死罪之名。其实皆以为善, 为之不知其义, 被之空言而不敢辞。" It means "As a minister or a son, if he doesn't understand the principles in *the Spring and Autumn Annals*, he will be condemned to death for plotting to usurp the throne and killing the king or his father. In fact, they both think they are doing something good, but do it without knowing why, and suffer unfounded criticism without courage to refute it".

"Melamine Incident", the "Changsheng Vaccine Incident", and the "Xu Yuyu Incident" were all caused by the deviation in science and technology, and the improper use of their achievements.

"No rose can bloom on the iceberg. Similarly, science can't easily bear fruit that can be regenerated in a socio-cultural environment that is not conducive to scientific development. For science to develop smoothly, the socio-cultural environment must be integrated with it. Science is truth-seeking. If the cultural environment in which it's located relies on fakeness to achieve aims, then the science will just be like a ball of red charcoal thrown on the snow. How can it burn? If science is advocated while social myths are created, it's like one foot is moving forward, the other moving backward. How can one move? The socio-cultural environment in which science can develop well is the one in which the pursuit of truth is regarded as the basic value. Only when truth becomes the attitude held by the majority of intellectuals in a social culture can science be truly supported" (Haiguang 1988). It can be seen that humanity is of great special significance to science. It's the "beacon" for scientific development, the "rectifier" for scientific ethics, and the "new philosophy" for "Tech for Social Good".

Technology is the key to the gates of both heaven and hell. Which door to open depends on the guidance of humanistic spirit. Only under the guidance of humanistic spirit can science and technology advance toward what is most conducive to the good development of mankind and the tech for social good.[37] The great philosopher of the Ming Dynasty Wang Yangming said, "Everyone has a conscience". He added, "The knowledge of good and evil is conscience". Conscience is the innate ability to recognize good and evil. Goodness is the ability and effort to fulfill oneself and others; to fulfill the world; and to bring more beauty, love, and light into the world. "Tech for Social Good" is the pursuit of science full of humanistic care and humanity full of scientific wisdom based on human being's free existence and development and liberation. Its proposition means humans have a further understanding of the relationship between people and science and technology. "Technology is an ability. Being good is a choice".[38] "Be people-oriented, altruistic and good". Under the guidance of altruism and sharing culture, science and culture will be blended and be independent of each other with the aid of sovereignty blockchain. They will be harmonious but different while maintaining appropriate tension. Humans live

---

[37] In May 2019, McKinsey released a report ("Tech for good: Smoothing disruption, improving well-being"), proposed the concept of "tech for social good", and pointed out that technology itself, as a tool, may have some negative impact in a short period of time, but governments, business leaders, and individuals will ensure that new technologies have a positive impact on society. In November 2019, on 21st anniversary of Tencent's founding, Tencent officially announced its new Tencent Culture 3.0. Tencent's vision was upgraded to "user-oriented and tech for social good", which means everything is based on user value, and social responsibility will be integrated into products and services, while promoting technological innovation and cultural heritage to help upgrade all industries and promote sustainable social development. Tencent, as the advocate and practitioner of "tech for social good", believes that "avoiding the evils of technology and realizing technology for good" is "tech for social good".

[38] Ma Huateng. User-oriented and Tech for Social Good—Written at the launch of Tencent Culture 3.0. QQ.com, 2019, https://tech.qq.com/a/20191111/007014.htm.

in harmony with nature and society, with the proper coexistence of animate and inanimate bodies.

5.3.3 Cultural Significance of Yangming Philosophy in Building a Community with a Shared Future for Humanity

Peace and development remain the theme of our era. Meanwhile, instability and uncertainty are more acute, and humans are facing common challenges. The global governance model and pattern formed after the Second World War are unsustainable due to the "new isolationism" pursued by the United States and the impact of the Brexit (Zhengzhong 2019). The eastern civilization represented by China has moved toward the center of the world stage. "A Community with a Shared Future for Humanity" proposed by China has increasingly become a common norm of values to the reform of global governance system and the construction of a new type of international relations and international order. China's solutions and wisdom are leading the way for a new global governance order.[39]

5.3.3.1 Troubles of the World and Rule of China

Shift of Civilization Center: from the Decline of Western Civilization to the Revival of Eastern Civilization. In the past 2000 years, the center of human civilization has undergone two major adjustments. The first took place in the 1860s. After occupying a leading position in world GDP for over 1,800 years, China was overtaken by European nations after it missed the opportunity of the first industrial revolution due to its adherence to the small-scale peasant economy and its policy of "self-seclusion". The center of world politics, economy, and culture (the center of world civilization) began to shift from the China-centered East to Europe.[40] In twentieth century, the center of world civilization underwent a second adjustment from Europe to the United States. After the two world wars, the European economy was blown hard. The European powers that used to dominate the world experienced declining economy. However, the United States, which was only a British colony in America, took advantage of its inherent "island advantage" of being far away from the main European battlefields and made war profits by selling materials and weapons

---

[39]Feng Yanli, Tang Qing. The Profound Connotation and Times Value of a Community with a Shared Future for Mankind. People's Daily Online. 2017. https://theory.people.com.cn/n1/2017/1212/c40531-29702035.html.

[40]According to statistics, by 1900, one-fifth of the world's land and one-tenth of its population had been carved up by European powers, and European civilization had overtaken the globe.

and providing war loans to the warring nations. At the end of the Second World War, the United States rose directly from a world power to a superpower. With its GDP accounting for more than 60% of the world's total, it naturally became the world's new economic and cultural center. Since then, Western civilization, with the United States as its core, swept the globe. Committed to "freedom, democracy and human rights" as "universal values", the US wantonly interfered in the internal affairs of other nations under the banner of "democracy without borders" and "human rights above sovereignty", advocated hegemonism, and subjectively assumed that those standards in the West, e.g., democratic elections, multiparty system, and the separation of powers, are the ideal model of modern political civilization, while suppressing those socialist nations that deviate from its values.[41] It can be said that "universal values" have become another manifestation of Western "civilization arrogance". With the gradual rise of emerging developing countries, such as China, India, and Brazil in the end of twentieth century, division and combination of various international forces accelerated, and relations between major powers once again entered a new stage of all-round wrestling. The world is facing "unprecedented changes in a century", and a new round of major adjustments in the center of world civilization is in the making. The cultural influence of emerging market nations and developing countries represented by China is gradually showing an upward momentum, and China is approaching the center of the world stage.

Chaos in the West: a Major Source of World Economic Fluctuation and Social Unrest. Around the 1990s, when the collapse of the Soviet Union and the upheaval of Eastern Europe delighted some Westerners, there was a buzz about the "end of history". But shortly after the beginning of twenty-first century, there were a lot of problems about the fate of the West. The subprime mortgage crisis in the United States and the international financial crisis it triggered, the European debt crisis, the Brexit, the failure of the Italian referendum to amend the constitution, the European refugee crisis, etc., coupled with social class confrontation, the spread of isolationism, the growth of populism, surges of right-wing extremism, frequent terrorist attacks, "black swan" incidents in elections, social protests and riots caused by racial discrimination, etc. These "government deficits" hurt the Western society badly (Ming 2017). In addition, the West also faced institutional crisis, democratic crisis, cultural crisis, etc. These Western "chaotic scenes" were not independent of each other, but with internal relations. Their interaction helped form a "chaos chain" or "chaos group", and eventually plunged the Western democratic model into a serious crisis and severe challenges. The chaos in the West is neither a single phenomenon nor an accidental phenomenon, but a normal one that occurs in multiple dimensions, fields, and planes, characterized by lasting time, wide space, and far-reaching influence (Gang and Lianfang 2019). The chaos in the West shows the "rule

---

[41] In fact, the so-called democracy is a peaceful compromise mechanism established on the basis of "universal values", namely, to express demands through one man, one vote, so people come to a compromise with different position, and such a democracy, once linked to benefits, will be a fuse that scatters national development power, weakens national identity, changes internal chaotic into the tyranny of the majority, and even causes large-scale riots and even war.

of the West" is experiencing a systemic crisis and has become a major source of world insecurity and instability, affecting world peace and development seriously.

Rule of China: Embarking on the Road to Great Rejuvenation and Setting a New Course for Global Governance. China's path of peaceful development is in sharp contrast to the chaos in the West. More than 70 years after its founding, China is developing quickly in a way that the West doesn't recognize. An old China with chronic poverty has become an increasingly prosperous socialist new China with powerful economy, science and technology, national defense as well as comprehensive national strength in the forefront of the world. "Two Miracles" of rapid economic development[42] and long-term social stability[43] rarely seen worldwide shock the West and the whole world. These "two miracles" complement and supplement each other, practically demonstrating of the power of the socialist system with Chinese characteristics (Hui 2020). It fully shows the socialist system with Chinese characteristics and the national governance system guided by Marxism, rooted in the Chinese land, have a profound Chinese cultural foundation and win the strong support of the people. It also shows the socialist system with Chinese characteristics and the national governance system are the institution and governance system with great vitality and superiority that can sustain the progress of a nation with a population of nearly 1.4 billion and ensure the Chinese nation, with a history of 5,000-odd years of civilization, achieves the "Two Centenary Goals", and then great rejuvenation.[44] "China's rise is not only an economic event but also a cultural one", says Michael Barr, an American scholar. The rise of great powers has fully proved that economic development represents

---

[42]The "miracle of rapid economic development" is mainly reflected in the following aspects: more than 70 years since the founding of New China, especially in the new period of reform and opening up, and since the 18th Chinese National Congress of the Communist Party of China, China has completed in a few decades the industrialization that had taken developed countries centuries to complete, and its social productivity has been greatly liberated and developed. China's economic strength and comprehensive national power have been significantly enhanced. China has become the second largest economy, the largest manufacturer, the largest trader of goods, the largest holder of foreign exchange reserves, and the second largest destination and source of foreign direct investment in the world. With reform and opening up in more than 40 years, China's economic growth has contributed about 18% to world economic growth on average every year, and in recent years about 30%.

[43]The "miracle of long-term social stability" is mainly reflected in the following aspects: over the past 70 years since the founding of the People's Republic of China, especially in the new period of reform and opening up and since the 18th National Congress of the Communist Party of China, China has undergone tremendous economic and social changes as well as several major tests. Before the reform and opening up, China had withstood the War to Resist US Aggression and Aid Korea, the 3 years of natural disasters, the "Cultural Revolution", and the strong earthquakes in Tangshan and Fengnan, Hebei Province. After reform and opening up, China has again endured such major tests as the political turmoil in spring and summer of 1989, the Asian financial crisis of 1997, the devastating floods of 1998, the bombing of Chinese Embassy in Federal Republic of Yugoslavia by the United States-led NATO in 1999, the SARS in 2003, devastating Wenchuan earthquake in Sichuan and international financial crisis in 2008 as well as COVID-19 in 2020.

[44]Xinhua News Agency. Decision of the Central Committee of the Communist Party of China on Several Major Issues Concerning the Adherence to and Improvement of the System of Socialism with Chinese Characteristics and the Promoting of Modernization of the State Governance System and Governance Capacity. *People's Daily*, November 6, 2019, p. 1.

not only the enhancement of "hard power", but also the synchronous enhancement of "soft power". And it is the "rule of China" that provides the world with great stability at a time when the undercurrents of "anti-globalization" are surging.[45] As a responsible major nation and a contributor to global governance system, China has long gone beyond its focus on self-development and construction. It has taken the initiative to shoulder international responsibilities, contribute its wisdom, and offer its own solutions to the problems worldwide.

5.3.3.2   Cultural Connotation of the "Rule of Conscience"

"Rule of China", originated from Confucianism, has its core in the "rule of conscience". The "rule of conscience" is to integrate Yangming Philosophy with modern governance, to achieve coordination and balance in strengthening moral rationality with conscience as the core, so as to realize the goal of building a Community with a Shared Future for Humanity. The essence of the "rule of conscience" is to build an orderly, fair, dynamic, upward, and prosperous society, which is the social ideal of "all things are one unity with benevolence" advocated by Wang Yangming, and its cultural connotation is the cultural value system constructed by "the mind is the principle", "the unity of knowledge and action", and "the extension of conscience".

Mind is the Principle: Theoretical Basis of the "Rule of Conscience". The proposition that "the mind is the principle" existed in ancient times. It was only until Wang Yangming proposed the theory that it represented the awakening of individual subject consciousness. Wang Yangming believed that "mind" was the individual mind, existing transcendentally in every human's subjective consciousness in the form of conscience, as the "Four Vitures" illustrated by Mencius.[46] In addition, Wang Yangming believed "my mind is the principle of the universe", meaning my mind and all things were of one unity. All things were in my mind, and the existence of my mind was inseparable from all things. That is to say, the world of each person was the world created by his own mind to a large extent, and the meaning of this world was also endowed by his own "mind". What kind of "heart" would lead to what kind of "world". Therefore, Yangming Philosophy first determined the connotation of "the mind is the principle", that is, "the mind is inseparable from the principle and things", with its significance in the emphasis on human moral subjectivity and human value, which is also the starting point of Yangming Philosophy.

Unity of Knowledge and Action: Theoretical Subject of the "Rule of Conscience". Wang Yangming believed "knowledge is the beginning of action, and action is the completion of knowledge" and "knowledge is the idea of action, and

---

[45] [Brazil] Oliver Stuenkel. Rule of China and the Future of the World. *Study Times*, January 15, 2018, p. 2.

[46] "Four Ends" refers to the four virtues that Confucianism believes are due, that is, "the heart of compassion is the end of benevolence; the heart of shame is the end of righteousness; the heart of resignation is the end of courtesy; and the heart of right and wrong is the end of wisdom". "Four Ends" is the major part of Mencius' thought, and his important contribution to the pre-Qin Confucianism theory as well.

action is the effort of knowledge". In other words, knowledge and action were consistent as they were rooted in the same noumenon. Therefore, "the unity of knowledge and action" was significant in that it completely brought the way of thinking that separated the theory and practice originated from ancient Greek philosophy. From the perspective of moral cognition, "the unity of knowledge and action" meant the unity of moral cognition and moral practice. "Knowledge" was the inner moral cognition of humans, while "action" was the outer behavior of humans. What Wang Yangming emphasized was to unify inner moral cognition and outer moral behavior. Hence, the significance of "the unity of knowledge and action" was to prevent people's "bad idea in a moment". When people were about to have "bad ideas" against moral and ethical codes, it is necessary to nip them in the bud so as to prevent the "bad idea" from lurking in people's thoughts and slowly growing. It can be seen that Wang Yangming's concept of "the unity of knowledge and action" was a process from knowing good to doing good, requiring people to put their ethical and moral knowledge into practice, so as to improve their moral personality. As "the motivation of goodness is just the beginning of goodness, not the completion of goodness. The goodness of thought is not really good until it is put into practice". Morality, serving as an invisible shackle at any time, not only blocks human freedom, but also ensures human freedom. Thus, "the unity of knowledge and action" became the criterion of Confucian moral image, the core of Yangming Philosophy, and the theoretical subject of "rule of conscience".

Extension of Conscience: Theory Sublimation of the "Rule of Conscience". "Extension of Conscience", the fundamental aim of Yangming Thought, marked the final analysis of Yangming Philosophy and fundamentally reshaped the structure of Confucianism. Neo-Confucianists used to believe "obtaining knowledge lies in the investigation of things", meaning we must start from the investigating of things for the purpose of obtaining knowledge. Wang Yangming invented a new way to integrate the "knowledge obtaining" in *The Great Learning* with Mencius's "conscience" theory. In his opinion, "conscience", inherent in humans, could help people "know good and evil" and make correct evaluation of their own behavior, guide their behavior choices, and urge them to abandon evil and turn to good. Therefore, "conscience" was a transcendental and universal moral consciousness in the form of knowledge of right and wrong. Through people's self-knowledge of "conscience", "the extension of conscience" was to help people "perceive" the "material desire" and "private interests" were the main causes of their blind "conscience", thus cultivating a kind of moral conscious initiative, to maintain or restore the nature—"Complete tolerance and calmness" of "conscience in my heart". That is to say, "the extension of conscience" emphasizes self-denial and selfishness to achieve fairness and justice. In addition, the core idea of "the extension of conscience" includes a concept of putting oneself in the place of others, that is, extending one's feelings outward, from the near to the far. The core idea of conscience is the way of loyalty and consideration, which is benevolence. "Loyalty" is doing one's best and "consideration" is putting oneself in the place of others. The practice of loyalty and consideration is to gradually extend outward from the individual's subjectivity, to make people evolve gradually from an individual to a unity with the universe. Wang Yangming said "Wind, rain, dew, thunder, the sun, the moon, stars, animals, plants, mountains and rocks are one with humans". He believed that human spirit is the link between human and heaven, earth, ghosts, and gods, so the human heart was "one body with heaven, earth, ghosts and gods". The start and close of human conscience corresponded to the shift of day and night of nature, so the human heart was one body with heaven and earth. "Benevolence" was given to all things, so all things were one with "benevolence". Therefore, Yangming Philosophy of "unity of all things" was based on morality, that is, conscience. It can be seen from this Yangming's philosophy was a compassion expanding outwardly, that is, from the individual to the family, to the society, to the ethnic group, to the whole of humanity, and even to all things worldwide. Therefore, "rule of conscience" is based on "conscience". That is to say, if "conscience" is lost, the "rule of conscience" will lose its spirit, and there will be no "rule of all things as one body". This is why Yangming promoted "all things are one unity with benevolence".

5.3.3.3 Contemporary Value and Future Significance of Yangming Philosophy to Global Governance

The world today is undergoing major development, transformation, and adjustment. Mankind is facing many common challenges, such as severe instability and uncertainty, lack of driving force for global economic growth, serious polarization between the rich and the poor, and the spread of threats such as terrorism, cybersecurity, and infectious diseases. Against this grim international backdrop, the responsibility of the Chinese civilization since ancient times is a Chinese solution to the inheritance of peace from one generation to another, to the momentum of development drive, and to shining civilization. Standing at the height of human history and world development, China proposes a Community with a Shared Future for Humanity, which is in line with the ideas of Marx and Engels. Marx and Engels once proposed that "only in the community can individuals obtain the means to develop their abilities in an all-round way, that is to say, only in the community can there be individual freedom",[47] and then they put forward the idea of "genuine community". "Genuine community" is the opposite of "false community". "Genuine community" means that communism is a union of free people and a society in which everyone develops freely and in an all-round way. In March 2013, President Xi Jinping proposed the concept of "a Community with a Shared Future for Humanity" to the world in his speech at the Moscow Institute of International Relations, pinpointing that human society today "is becoming more and more an interconnective community of shared future".[48] From the 2013 Annual Conference to the 2015 Annual Conference of

---

[47][Germany] Marx, [Germany] Engels. *Selected Works of Marx and Engels (Vol.1)*, translated by Central Compilation and Translation Bureau. People's Publishing House, 1995, p. 119.

[48]Zhang Minyan. Shared Future with Each Other, On a Community with a Shared Future for Humanity by Xi Jinping. Xinhuanet. 2019. https://www.xinhuanet.com/2019-05/07/c_1124463051.htm.

Boao Forum, the vision of a Community with a Shared Future for Humanity has leapt from "building a sense of a community with a Shared Future" to "moving towards a community with a Shared Future".[49] In 2017, "Building a Community with a Shared Future for Humanity" was listed into UN resolutions,[50] UN Security Council resolutions,[51] and UN Human Rights Council resolutions,[52] demonstrating China's important contribution to global governance. In the 19th Communist Party of China National Congress, President Xi Jinping mentioned the Community with a Shared Future for Humanity six times. Standing at the height of the progress of all humans, he made a solemn commitment to the world, "China will continue to play its role as a responsible major nation, take an active part in the reform and development of global governance system, and continue to contribute China's wisdom and strength".[53] Meanwhile, "a Community with a Shared Future for Humanity" was written into the *Constitution of the CPC*, which was amended and adopted at the 19th National Congress of the Communist Party of China, making this vision an unprecedented political height. In March 2018, the First Session of the Thirteenth National People's Congress voted to adopt the *Amendment to the Constitution of the People's Republic of China*, and "promoting the Community with a Shared Future for Humanity" was written into the preamble of the *Constitution*. This has brought "a Community with a Shared Future for Humanity" into China's legal system, marking it as a significant part of Xi Jinping Thought on Socialism with Chinese Characteristics for a New Era.[54]

A Community with a Shared Future for Humanity is at the core of the principle of joint consultation, contribution, and sharing of global governance. Its essence is a "global outlook" beyond the ideology of nation states, with the ultimate goal of building an "open, inclusive, clean and beautiful world of lasting peace, universal security and co-prosperity". This is a new form of human interdependence that takes economy, politics, and ecology as the link and transcends regions, nationalities,

---

[49]Zhu Shuyuan, Xie Lei. What is the Diplomatic Concept of "a Community of a Shared Future" Frequently Mentioned by Xi Jinping? People's Daily Online. 2015. https://cpc.people.com.cn/xuexi/n/2015/0610/c385474-27133972.html.

[50]Liu Gefei. "Building a Community of a Shared Future for Humanity" was First Written into the UN Resolution. Xinhuanet. 2017. https://www.xinhuanet.com//world/2017-02/12/c_129476297.htm.

[51]Liu Xiaodong. Inclusion of the Concept of Building a Community with a Shared Future for Humanity in Security Council Resolutions is a Major Contribution of China's diplomacy. Xinhuanet. 2017. https://www.xinhuanet.com/world/2017-03/23/c_1120683832.htm.

[52]Tang Lan. The Major Concept of a Community with a Shared Future for Humanity is Included in the UN Human Rights Council Resolution for the First Time. Xinhuanet, 2017, https://www.xinhuanet.com/world/2017-03/24/c_129517029.htm.

[53]Xi Jinping. Winning the Victory of Building a Well-off Society in an All-round Way and the Great Victory of Socialism with Chinese Characteristics in the New Era-Report on 19th National Congress of the Communist Party of China. Xinhuanet. 2017. https://www.xinhuanet.com/politics/19cpcnc/2017-10/27/c_1121867529.htm.

[54]Li Shenming. Historical Status and Global Significance of Xi Jinping Thought on Socialism with Chinese Characteristics for a New Era. QStheory.cn. 2017. https://www.qstheory.cn/dukan/qs/2017-12/31/c_1122175320.htm.

and nations. It is a common prerequisite for the development of human civilization. Therefore, at the critical moment of insufficient global growth energy, lagging global economic governance, and global development imbalances, China's proposal to a Community with a Shared Future for Humanity with the demeanor and ambition of a great power is an in-depth concern for people of all ethnic groups worldwide and an important manifestation of the responsibility of a major power. When outlining the basic principles for a Community with a Shared Future for Humanity, President Xi proposed that partnerships should "treat each other as equals and engage in mutual understanding and consultation", and that "civilization exchanges should be harmonious and inclusive in diversity" and ecosystems should "respect nature and develop in a way friendly to environment". The ideas of "cooperation", "win–win", and "universal benefit" in it coincide with the ideas of "peace, benevolence, and all under heaven of one family" in the essence of Chinese culture, showcasing the wisdom and pattern of "harmony being precious", "tolerance bringing respect", and "harmony in diversity" in traditional Chinese culture, exemplifying China's political philosophy of "the whole world as one community", "harmony of all nations", and "global peace". Tracing the history of Chinese civilization, we may find that the cosmological concept of "harmony between humans and nature", the international concept of "harmony of all nations", the social concept of "harmony in diversity", and the moral concept of "a kind heart" have formed in the Chinese civilization for more than 5,000 years. In the ideal universe of Confucianism, there are no different nations and borders or boundaries between nations and cultures. Confucianism pursues the unity of the world, and its fundamental value is cosmopolitan and universal. The Confucian cosmopolitanism holds that "all people of the world are brothers", which meets civilization in a pluralistic world. Among them, the Yangming Philosophy founded by Wang Yangming, a thinker and philosopher of the Ming Dynasty, was a collection of the essence of Confucian culture. Moreover, it advocated "all things are one unity", "the mind is the principle", "the unity of knowledge and action", and "the extension of conscience". Yangming Philosophy embraced the concern for all things, among which the innermost individual consciousness was conscience.

The initiative of a Community with a Shared Future for Humanity is China's solution, wisdom, and contribution to global governance, with its focus on the integration and co-governance of diverse civilizations. It is a fact that world civilizations are diversified. How can different values coexist without excluding each other? How to realize "all living creatures growing together without harming one another; ways running parallel without interfering with one another"? We are inspired by the conscience in Yangming Philosophy. Conscience, an important basis and organization for individual moral consciousness and moral choice as well as an internalized form of universal etiquette and *Tao* (law of universe), provides moral guidance for people's behavior. Although different civilizations take different forms, the pursuit of conscience is interlinked. Wang Yangming said, "If a person's mind has reached a pure and clear state, he will realize 'the unity of all things' and attain the realm of benevolence. Therefore, his spirit flows constantly, his ambition insightful, and there is no difference between himself and others, between himself and external things". "The heart of a sage is integrated with all things worldwide. He sees all people under heaven as equals regardless of kinship or distance. As long as they have courage, they are his brothers and children. The sage wants to educate them so as to realize his wish of unity of all things". "So, while loving my father, I also love the father of others and the father of everyone worldwide...The same is true for mountains, rivers, gods, birds, beasts, plants, and trees. I have loved them truly, so as to achieve my benevolence of all things. By doing so, my light and virtue will not be unrevealed, so I will truly be one with the universe". Wang Yangming advocated "the conscience of my heart should be applied to everything" and "all things are one unity with benevolence". He highlighted we should take conscience as the guide, keep the world in mind, have a sense of responsibility and a kind heart to others and groups, thus establishing a moral order universally recognized by the world and making the whole society a harmonious form. The idea it contains is to recognize and embrace the differences and diversity of civilizations, which is of great cultural significance to the initiative of a Community with a Shared Future for Humanity. In particular, the concern for the world and the essence of conscience contained in the philosophy of "all things are of one unity" constitute an aspect of the recognition and understanding of the world toward a Community with a Shared Future for Humanity, and offer effective assistance for all nations to recognize, accept, and identify with such a community. It can be said that Yangming Thought, especially Yangming Philosophy, is one of the cultural sources of a Community with a Shared Future for Humanity and the basic cultural connotation of the "rule of conscience". In the future, with the shift of human civilization center, Yangming Philosophy will be disseminated and popularized worldwide, offering more cultural nourishment and guidance for a Community with a Shared Future for Humanity. Eastern civilization will surely shine with a brighter light of conscience.

References

Comte I Auguste. *System of Positive Polity* (2 vols.) London: Longmans, Green & Co. 1875, pp. 566-567.

Ma Baobin et al. *Theory and Practice of Public Governance.* Social Sciences Academic Press, 2013.

Zhang Bencai. Outline of Future Law. *Law Science*, 2019, No. 7.

Hu Bo. The Sharing Model and the Future Development of Intellectual Property—Review on the Alternative Model of Intellectual Property. *Law and Social Development*, 2013, No. 4.

Chen Caihong. Artificial Intelligence and the Future of Humans. *Bookstore*, 2018, No.12.

China Blockchain Technology and Industrial Development Forum (CBD-Forum): White paper on blockchain technology and application development in China (2016). Official Website of CBD-Forum, technology and industry development in 2016.

Li Chuanjun, Order Change and Model Construction of Global Governance in Cyberspace, *Journal of Wuhan University of Science and Technology (Social Science Edition)*, 2019, No. 1.

Darwin. *The Descent of Man*, translated by Pan Guangdan and Hu Shouwen. The Commercial Press, 1997.

Lu Di. Online Video and Information "Communism". *News and Writing*, 2014, No. 1.

Chen Duan. The Modernization of National Governance by Digital Governance. *Frontline*, 2019, No. 9.

Xie Fang. Science Fiction, Futurology and Future Times. *Chinese Social Science*, January 25, 2013, p. A5.

Fan Feng. Theoretical Basis and Practical Path of the Community of Common Destiny for Cyberspace. *Journal of Hebei University (Philosophy and Social Science)*. 2018, No. 6.

[UK] Niall Ferguson. *Civilization*, translated by Zeng Xianming et al. China CITIC Press, 2012.

Wang Gang, Zhou Lianfang. The Western Disorder and Its Challenges and Enlightenments to China. *Leading Journal of Ideological & Theoretical Education*, 2019, No. 3.

[South Africa] Ian Goldin, [Canada] Chris Kutarna. *Age of Discovery: 21st Century Risk Guide*, translated by Li Guo. China CITIC Press, 2017.

Wang Guanyi. Philosophy and Social Science: Giving Play to the Guiding Function. *People's Daily*, August 4, 2013, p. 5.

Yin Haiguang. *Reappraisal of Cultural Change in Modern China*. China Peace Publishing House, 1988.

Wu Handong. Institutional Arrangements and Legal Regulation in Age of Artificial Intelligence. *Science of Law*, 2017, No. 5.

[Israel] Yuval Noah Harari. *Homo Deus: A Brief History of Tomorrow*, translated by Lin Junhong. China CITIC Press, 2017.

[Israel] Yuval Noah Harari. *21 Lessons for 21st Century, translated by Lin Junhong*. China CITIC Press, 2018.

[US] James Hendler, [US] Alice M. Mulvehill. *Social Machines: The Coming Collision of Artificial Intelligence, Social Networking, and Humanity*, translated by Wang Xiao et al. China Machine Press, 2018.

[US] A.J. Herschel. *Who is Man*, translated by Wei Renlian. Guizhou People's Publishing House, 1994.

Ni Hongbo. Safe Use of Nuclear Force. *The Science News*, 2017, No. 6.

Yu Hua. University Is the Home for the Co-Prosperity of Science and Philosophy. *Legal System and Society*, 2019, No. 32.

Li Hui. Artificial Intelligence: The Technology Wave That Changes the World. *Information Security and Communications Privacy*, 2016, No. 12.

Jiang Hui. Giving Full Play to the Advantages of the System and Achieving "Rule of China" Successfully. *People's Daily*, January 7, 2020, p.10.

Dong Hui, Li Jiali. Path Choice of Network Governance in the New Era: A Community with Shared Future in cyberspace. *Study and Practice*, 2017, No. 12.

Wang Jianmei. The Four Dimensions of a Community of a Shared Future for Cyberspace. *China's Collective Economy*, 2019, No. 25.

Chen Jie, Fang Yiyun. "Black Swan" and "Gray Rhino". *Financial Times*, September 8, 2017, p.10.

Liu Jinchang. *Artificial Intelligence Changes the World: Robots Towards Society.*. China Water Resources and Hydropower Publishing House, 2017.

Xi Jinping. Speech at the Symposium on Philosophy and Social Science. Xinhuanet. 2016. http://www.xinhuanet.com//politics/2016-05/18/c_1118891128.htm.

Xi Jinping. Winning the Victory of Building a Well-off Society in an All-round Way and Winning the Great Victory of Socialism with Chinese Characteristics in the New Era—Report on 19th National Congress of the Communist Party of China. *People's Daily Online*. 2017. http://cpc.peo ple.com.cn/n1/2017/1028/c64094-29613660-14.html.

Xi Jinping. Speech at the 19th Academician Conference of Chinese Academy of Sciences and the 14th Academician Conference of Chinese Academy of Engineering. xinhuanet. 2018. http://www.xinhuanet.com/politics/2018-05/28/c_1122901308.htm.

Li Kaifu. *AI-Future*. Zhejiang People's Publishing House, 2018.

Zhao Kejin. The Change of International Order and China's Role in the World. *People's Tribune*. 2017, No. 14.

Key Laboratory of Big Data Strategy. *Block Data 2.0: Paradigm Revolution in the Era of Big Data*. CITIC Press, 2016.

Key Laboratory of Big Data Strategy. *Block Data 3.0: Orderly Internet and Sovereignty Blockchain.* CITIC Press, 2017.

Key Laboratory of Big Data Strategy. *Data Rights Law 1.0: Theoretical Basis of Digital Power.* Social Sciences Academic Press, 2018.

[US] Henry Kissinger. *World Order*, translated by Hu Liping, Lin Hua, Cao Aiju. China CITIC Press, 2015.

Long Kun, Zhu Qichao. Rule-making for Cyberspace: Consensus and Discord. *Global Review*, 2019, No. 3.

[US] Ray Kurzweil. *The Singularity is Near: When Humans Transcend Biology*, translated by Li Qingcheng et al. China Machine Press, 2011.

Xu Lei. The Third Wave of Artificial Intelligence and Some Cognitions. *Science (Shanghai)*, 2017, No.3.

[Russia] Lenin. *Selected Works of Lenin (Vol. 4)*, translated by The Compilation and Translation Bureau of the Works of Marx, Engels, Lenin, Stalin. People's Publishing House, 1972.

Ma Lijuan. Research on Governance Theory and A Review of Its Value. *Journal of Liaoning Academy of Governance*, 2012, No. 10.

[German] Marx, [German] Engels. *Marx and Engels Collected Works (Volume 1), (Volume 3)*, translated by the Central Compilation and Translation Bureau of the Works of Marx, Engels, Lenin and Stalin. People's Publishing House, 1972.

Xin Ming. Ignoring the Causes of Western Chaos. *People's Daily*, July 16, 2017, p. 5.

Wang Mingjin. The Future of Global Cyberspace Governance: Sovereignty, Competition, and Consensus. *People's Tribune: Frontiers*, 2016, No. 4.

Nagel T. The *The Possibility of Altruism.* Princeton: Princeton University Press. 1978, p.3.

[US] Nassim Nicholas Taleb. *The Black Swan: The Impact of The Highly Improbable*, translated by Wan Dan. China CITIC Press, 2011.

Wu Ning, Zhang Shujun. On Internet and Communism. *Journal of Changsha University of Science and Technology (Social Science edition)*, 2018, No. 2.

Pan, Wang and Xiao Sisi, Zhou Ying. 2019. Focus on the "Gene-edited Baby" Case. In *People's Daily*, 11.

Liu Qi. Beware of Technological Evil in the Age of Technology. *China Development Observation*, 2019, No. 15.

Feng Qingqing. Standing at The Tide of The Times, Giving the Leading Voice of Thinking: Shouldering the Responsibility and Mission of Prospering Philosophy and Social Science. *Hunan Daily*, July 7, 2016, p. 8.

[UK] Adam Smith. *The Theory of Moral Sentiments*, translated by Xie Zonglin. Central Compilation & Translation Press, 2008.

[UK] Adam Smith. *The Theory of Moral Sentiments*, translated by Jiang Ziqiang et al., The Commercial Press, 2015.

Wu Song, Li Yaqian. The New Meaning of "Black Swan" and "Gray Rhino". *Chinese Learning*, 2017, No. 11.

Ye Suibing. On the Operation Rules of a Community of a Shared Future for Cyberspace. *Economic and Social Development*, 2018, No. 3.

[US] Alvin Toffler. *The Third Wave*, translated by Huang Mingjian. China CITIC Press, 2018.

[Australia] Toby Walsh. *Will Artificial Intelligence Replace Human Beings?* translated by Lu Jia. Beijing United Publishing Co., Ltd., 2018.

Ma Wei. In Addition to the Black Swan, You Need to Know the Gray Rhino—After Reading "The Gray Rhino". *China Entrepreneur*, 2017, No.7.

Shen Weixing. Highlighting the Legal System Construction of Digital Economy in the Implementation of Big Data Strategy. *Guangming Daily*, July 23, 2018, p. 11.

Jiang Wenfu. Life Culture: Harmony Between Science and Humanity. *Guangming Daily*, February 17, 2016, p. 14.

Zheng Wenming. China's Choice of Internet Governance Model. *China Social Science Daily*, August 17, 2017, p. 3.

[Germany] Wilhelm Windelband. *A History of Philosophy*, translated by Luo Daren. The Commercial Press, 1993.

World Economic Forum. "The global risks report 2019 (14th edition)". *World Economic Forum.* 2019. http://www3.weforum.org/docs/WEF_Global_Risks_Report_2019.pdf.

[US] Michele Wucker. *The Gray Rhino: How to Recognize and Act on the Obvious Danger We Ignore*, translated by Wang Liyun. China CITIC Press, 2017.

Gao Xiaoyan. The Prototype of Japanese Army Unit 731—Beiyinhe Bacteria Experiments. *Japanese Invasion of China History Research*, 2014, No. 1.

Li Xiao, Liu Junqi, Fan Mingxiang. Research on Prevention and Response Strategies of WannaCry Extortion Virus. *Computer Knowledge and Technology*, 2017, No. 19.

Fang Xingdong. Research on Prism Gate Incident and Global Cyberspace Security Strategy. *Modern Communication (Journal of Communication University of China)*, 2014, No. 1.

Zhao Xudong. The Impact of New Technological Revolution on National Sovereignty, *Europe*, 1997, No.6.

Chen Xuebin. Grey Rhino. *Heilongjiang Finance*, 2018, No. 2.

Sun Xiaoli. Science and Philosophy in 21st Century. *Expanding Horizons*, 2003, No. 6.

Qian Xuesen. *Letters of Qian Xuesen (Vol. 7)..* National Defense Industry Press, 2007

Yang Yanchao. *AI Law*. Law Press. China, 2019.

Li Yanhong. *Intelligent Revolution: Social*, Economic and Cultural Reforms Meeting the Era of Artificial Intelligence. China CITIC Press, 2017.

Xu Yaotong. "Modernization of National Governance" should be Mentioned. *Beijing Daily*, June 30, 2014, p.18.

Chen Yiming. Major Events of Iran Nuclear Issue. *People's Daily*, January 11, 2006, p. 3.

Fu Ying. International Order and China's Actions. *People's Daily*, February 15, 2016, p.5.

Liu Ying, Wu Ling. Global Cyberspace Governance: Chaos, Opportunity and China's Proposition. *Cognition and Practice*, 2019, No. 1.

Zheng Yongnian. The stage of passive response has passed. Experience shows that passive response, no matter how good it is, is far from enough—an effective response to the US right to define. *Beijing Daily*, September 2, 2019, p.16.

Cai Yuanpei Research Association of China. *The Complete Works of Cai Yuanpei (Vol. 2)*. Zhejiang Education Publishing House, 1997.

Chen Yugang. International Order and the View of International Order (Preface). *Fudan International Studies Review*, 2014, No. 1.

Lian Yuming. *Report on Guiyang Social Governance System and Governance Capacity Development*. Contemporary China Publishing House, 2014.

Lian Yuming. Salute to the New Era: The Application of Governance Technology based on Sovereignty Blockchain in Consultative democracy. *CPPCC Fortnightly*, 2018, No. 6.

Wang Yunling. "Natural Person" and "Technological Person": An Ethical Review on Gene-editing Baby Events. *Journal of Kunming University of Science and Technology (Social Science Edition)*, 2019, No. 2.

Liu Yuqing, Gong Yanli. Study on Security Threats and Countermeasures in Era of Cyber Warfare. *Information Research*, 2014, No. 11.

[Japan] Matsuo Yutaka. *Artificial Intelligence Frenzy: Will Robots Surpass Humans?* translated by Zhao Hanhong et al. China Machine Press, 2016.

Guo Zhangcheng. Five Singularities Possible to Artificial Intelligence. *Theoretical Vision*, 2018, No.6.

Zhi Zhenfeng. Global Vision and China's Responsibility for a Community of a Shared Future for Cyberspace. *Guangming Daily*, November 27, 2016, p. 6.

Xu Zhengzhong. Global Governance Innovation and Chinese Wisdom. *Study Times*, November 15, 2019, p. 2.

Li Zhifei. The Water Diplomacy: a New Proposal on the Periphery Safety Construction of China. *Academic Exploration*, 2013, No. 4.

Diao Zhiping. Reflecting on the Essence of Global Crisis from the Advantages and Disadvantages of Traditional Cultural Models. *China Soft Science*, 2003, No. 2.

Huang Zhixiong, Ying Yaohui. The Impact of the United States on International Law in Cyberspace and its Implications for China. *Fudan International Studies Review*, 2017, No. 2.

Li Zhiyong. *Ultimate Replication: How Artificial Intelligence Will Shape Our Society*. China Machine Press, 2016.

Postscript

On the last day of 2016, the Information Office of the Guiyang Municipal People's Government pioneered in releasing a local manifesto for the development of Blockchain named *Guiyang Blockchain Development and Application White Paper* which put forward the groundbreaking concept of sovereignty blockchain. Soon afterward, sovereignty blockchain was selected as one of the "Top 10 New Words for Big Data" by China National Committee for Terms in Sciences and Technologies and released at the 2017 China International Big Data Industry Expo, making it a formal sci-tech term in China.

The Key Laboratory of Big Data Strategy, jointly built by Guiyang Municipal People's Government and the Beijing Municipal Science and Technology Commission, is an interdisciplinary, professional, international, and open research platform as well as a new high-end think tank for big data development in China. Since 2015, it has launched a "Governance Technology Trilogy" consisting of block data, digital rights law, and sovereignty blockchain, which are hailed as the three pillars for reconstructing a new era of digital civilization and have exerted great impact worldwide. The concept of sovereignty blockchain has been our major research object, on which we have already launched three books. *Block Data 3.0.* subtitled *Orderly Internet and Sovereignty Blockchain*, focuses on technology-based to institution-based governance approach. *Redefining Big Data* devotes a chapter to the significance of sovereignty blockchain to innovate organization, governance system, and operating rules. *Data Dictionary* has developed a system of knowledge and terminology with sovereignty blockchain as an important component, and has gained recognition and recommendation from UNESCO International Knowledge Centre for Engineering Science and Technology (IKCEST).

Sovereignty blockchain 1.0: Orderly Internet and Community with a Shared Future for Humanity is a major fruit of the Key Laboratory of Big Data Strategy based on theoretical research of block data and data rights law. It also proactively responds to President Xi's major speech spirit and proposition to "make our nation to take the lead in terms of theory, innovation and industry in the emerging field of blockchain". Firstly, the book elaborates the basic law of Internet development from the Internet of Information to the Internet of value, then to the Orderly Internet;

© Zhejiang University Press 2021
L. Yuming, *Sovereignty Blockchain 1.0*,
https://doi.org/10.1007/978-981-16-0757-8

secondly, it categories data sovereignty theory, social trust theory, and smart contract theory as the three new theories of the era; thirdly, it discusses the significance of Tech for Social Good and Wang Yangming's Philosophy in building a Community with a Shared Future for Humanity. The book is a study of blockchain based on the Orderly Internet and a Community with a Shared Future for Humanity with the hope of providing new perspectives, concepts, and ideas for the development and application of blockchain. In the future, we will publish a series of theoretical monographs on sovereignty blockchain, whose themes include new forces to change the world of tomorrow, digital government leading the way, consultative democracy changing the world, and Chinese wisdom in global governance. The book series will provide Chinese solutions to global governance of Internet and contribute Chinese wisdom in building a Community with a Shared Future for Cyberspace.

The Key Laboratory of Big Data Strategy was responsible for organizing seminars, carrying out in-depth research, and the writing of the book. Lian Yuming presented the general approach and core ideas of the book, as well as the overall design of its framework. Long Rongyuan and Zhang Longxiang refined the outline and thematic ideas. Lian Yuming, Zhu Yinghui, Song Qing, Wu Jianzhong, Zhang Tao, Long Rongyuan, Song Xixian, Zhang Longxiang, Zou Tao, Chen Wei, Shen Xudong, Yang Lu, and Yang Zhou were responsible for writing the book, among whom Long Rongyuan and Zhang Longxiang were responsible for the final compilation and editing. Chen Gang offered many important forward-looking and instructive points for this book. Zhao Deming, Member of the Standing Committee of the Guizhou Provincial CPC Committee and Secretary of the Guiyang Municipal CPC Committee; Chen Yan, Deputy Secretary of the Guiyang Municipal CPC Committee and Mayor of Guiyang; Xu Hao, Member of Guiyang Municipal CPC Committee Standing Committee and Executive Deputy Mayor of Guiyang; and Liu Benli, Member of the Standing Committee of the Guizhou Provincial CPC Committee and Secretary General of the Municipal CPC Committee of Guiyang have all contributed a wealth of constructive thoughts and insights. After the completion of the manuscript, an academic seminar was held at the Key Laboratory of Big Data Strategy Research Base in Zhejiang University. Many experts and scholars exchanged opinions on respecting topics and offered many insights from a variety of perspectives, including Prof. Ben Shenglin, Dean of International Business School and Academy of Internet Finance of Zhejiang University; Yang Xiaohu, Prof. of College of Computer Science and Technology, Director of Blockchain Research Center, and Associate Dean of Academy of Internet Finance, Zhejiang University; Li Youxing, Prof. of Zhejiang University Guanghua Law School and Associate Dean of Academy of Internet Finance; Prof. Zhao Jun, Associate Dean of Zhejiang University Guanghua Law School; Zhang Ruidong, Tenured Professor at the University of Wisconsin-Auclair, Director of Blockchain Office at the Academy of Internet Finance, Zhejiang University; Prof. Zheng Xiaolin, Deputy Director of the Artificial Intelligence Institute of Zhejiang University; Chen Zongshi, Associate Professor at Department of Sociology of Zhejiang University; and Yang Lihong, Associate Professor of International Business School of Zhejiang University. It is appropriate to say that the book crystallizes the collective wisdom of the experts. Special thanks to the leaders and editors of Zhejiang University Press. President Lu Dongming, with his visionary thinking, unique judgment, and extraordinary courage, gave high recognition and support to the book and organized a number of editors to carefully plan, edit, and design it. Without their support, the book would not have made it to the readers as scheduled.

As the technology term becomes a term of the time, sovereignty blockchain may change the world in ways out of our imagination. If blockchain is the greatest human technological innovation of the early twenty-first century, then the development of the sovereignty blockchain will surely become the most exciting and anticipated native innovation of the second half of the twenty-first century. Such innovation is full ranged, covering not only technological innovation, but also theoretical, institutional, and mode innovation. The blockchain sketched human vision toward a digital civilization, while sovereignty blockchain provides a key that all of us are anticipating. This desired key will open the door to a future of digital civilization.

When the book is ready for publication, the world is at a critical juncture for the prevention and control of COVID-19. All nations are working around the clock at sci-tech research to give full play to technological achievement in combating the pandemic. A worldwide technological battle is under way as the Internet, big data, artificial intelligence, blockchain, and other new-generation information technologies have shown their strengths. It is foreseeable that after the pandemic, the new generation of information technology will not only be seen as a new drive of economic development, but also as a crucial prop for the modernization of governance system and capacity. The construction of blockchain, especially of sovereignty blockchain, will move from vision to reality with a gradual expansion and deep integration to all aspects of governance and services. Governance technology will become an important means of national governance modernization. We hope that some of our thoughts can provide reference for the application of governance technology, the innovation of governance systems, and the operation of governance scenarios. Blockchain is such a hot technological topic that it is constantly heating up, inspiring unanimous views and understandings from various sectors. In the compilation of the book, we have tried our best to collect the latest literature and absorb the latest views in order to enrich the ideas contained. However, there might be some inaccuracy in our views with our limited knowledge, ability, and cognition, as well as with various disciplines the book covers. Therefore, we are open to any criticism and suggestions from readers on any mistakes in the book, including references and sources. The liability is ours.

Key Laboratory of Big Data Strategy

March 10, 2020

Glossary

A A Community with a Shared Future for Cyberspace, A self-governing society, Absence, Acquaintance society, Activation dataology, Affinity, Algorithm hegemony, Algorithm trust, Algorithmic credit, Algorithmic discrimination, Alienation of sovereignty, Amalgamation, Artificial intelligence, Artificial Intelligence + Internet of Everything, Association and fusion, Atomic structure theory, Augmented Reality (AR), Automatic driving

B Big data makes a smarter world, Binary World, Bitcoin, Bitcoin World, Black Swan, Block data, Block data organization, Block data value chain, Blockchain, Blockchain evidence, Borderless society, Brain science, Brain-computer interface, Bureaucracy, Business relationship, Byzantine failures

C Central bank digital currency (CBDC)/(DC/EP), Centralization, Chaos Theory, Classical scientific paradigm, Cloud computing, Cognitive intelligence, Collective human rights, Collective maintenance, Communicative action theory, Community with a Shared Future for Humanity, Complex Giant System, Complex Science Paradigm, Complex theory, Complexity Science, Computational intelligence, Computational power, Consensus, Consensus Mechanism, Consensus Machine, Consensus Network, Consensus System, Consensus System, Contract, Contract-based trust, Contractual society, Correlation Analysis, Credibility, Credit mechanism, Credit society, Cross-link transmission, Cross-validation, Cross-border data flow, Cross-border payment, Cross-chain technology, Crowdfunding, Cryptology, Cyberattack, Cyberhegemonism, Cyberhegemony, Cybermilitarism, Cybersecurity, Cyberterrorism, Cybercrime, Cyberliberalism, Cyberspace

D Data, Data acquisition right, Data archiving right, Data asset investment, Data asset preservation, Data asset registration, Data authentication, Data barriers, Data boundary, Data cage, Data capital theory, Data capitalism, Data capitalization, Data capitalization, Data confidentiality, Data confirmation, Data congestion, Data correction right, Data credibility, Data deletion right, Data divide, Data empowerment, Data empowerment, Data evolution, Data flow, Data footprint, Data force, Data game theory, Data governance, Data hegemonism, Data independence, Data intelligence, Data jurisdiction, Data localization, Data mining,

© Zhejiang University Press 2021
L. Yuming, *Sovereignty Blockchain 1.0*,
https://doi.org/10.1007/978-981-16-0757-8

Data modification right, Data monopoly, Data multi-jurisdiction, Data open source, Data ownership, Data ownership system, Data person, Data personality right, Data philosophy, Data portability, Data potential difference, Data pricing, Data privacy, Data property, Data property right, Data protectionism, Data quality, Data relations, Data resourcing, Data revenue right, Data right, Data rights, Data rights law, Data search, Data security, Data separation, Dataset, Data sharing, Data isolation, Data sovereignty, Data sovereignty theory, Data space, Data standard, Data storage, Data subject, Data superpower, Data supply, Data technology, Data territory, Data terrorism, Data transactions, Data value, Dataism, Datalization, Decentralization, Decentralized Autonomous Corporation, Decentralized Autonomous Organization, Decentralized Autonomous Society, Deep learning, Delayering, Destratification, Detrust, Differential Mode of Association, Digital administration, Digital affluence, Digital affluent population, Digital blackmail, Digital certificate, Digital citizen, Digital civilization, Digital coercion, Digital consciousness, Digital contract, Digital currency, Digital divide, Digital economy, Digital elite, Digital exploitation, Digital extreme poverty population, Digital finance, Digital fraud, Digital future, Digital governance, Digital government, Digital human, Digital human rights, Digital identity, Digital identity chain, Digital labor, Digital literacy, Digital middle class, Digital migration, Digital object, Digital order, Digital planet, Digital poverty, Digital robbery, Digital signature, Digital society, Digital space, Digital survival, Digital system, Digital technique, Digital technology, Digital theft, Digital ticket, Digital trust, Digital trust model, Digital twin, Digital violence, Digital wallet, Digital world, Digitization, Disembedding, Disembedding Mechanism, Disorder, Distributed Database, Distributed Ledger, Distributed Storage, Distributed System, Division of Labor Era, Double spending, Driverless, Dual space

E Economic man, Edge Computing, Electronic bill, Electronic Money, Empathy, Encryption algorithm, Enigma, Enterprise data, Enterprise data sovereignty, Ethereum (ETH), Everyone-to-everyone society, Extraterritorial Jurisdiction, Federated Learning, Fifth Dimension

F Financial assets, Financial audit, Financial institution, Financial market, Financial service, Financial system, FinTech, Forced digital participation

G Game Theory, Gene editing, Gene sequencing, Gene-edited baby, Gene-edited man, Geopolitics, Ge-stell, Global commons, Global crisis, Global risk, Governance deficit, Governance efficiency, Governance of cyberspace, Governance technology, Grey Rhino, Hacker, Hash Collision, Hash Function, Heterogeneous society, High Reliability, Homo sapiens, Horizontal organization

H Hotspot reduction, Human ecosystem, Human rights, Human society space, Humanism, Hyperledger

I Identity, Identity chain, Immutability, Inclusive finance, Information asymmetry, Information distortion, Information flooding, Information Internet, Information plague, Information revolution, Information space, Information Technology, Informatization, Institution-based trust/institutional trust, Intelligentize, International Order, Internet fraud, Internet Governance, Internet MLM, Internet of

Everything, Internet of Things (IOT), Internet relationship, Internet Sovereignty, Interpersonal trust model, Isolated islands of information

K Kinship credit, Knowledge acquisition, Kondratieff cycle

L Legal system of data right, Legal Tender/Fiat Money, Libra, Long Tail Effect

M Machine consensus, Machine learning, Machine trust, Market consensus, Market economy, Market Failure, Maslow's hierarchy of needs theory, Means of production, Metage coin, Metcalfe's Law, Mind is Principle, Miner, Mobile Internet, Mobile payment, Model of social generative theory, Moral Economy, Motivation attribution, Multidimensional fusion, Multilateral Governance, Multilateralism, Multi-ownership of data, Multiple Ownership, Multistakeholder Governance, Mundellian Trilemma, Mutual Trust Mechanism

N National Interest, Natural person, Natural Rights, Negative human rights, Network domination, Network Monitoring, Network protocol, Network shape organization, Network trust, Network Trust Model, Network Virtual Property, Networking, Non-sovereignty power

O One Ownership for One Object, One thing one place, Open Access Movement, Open Government, Orderly internet, Overissuance of Money

P Paper contract, Paper currency/paper money, Peer-to-Peer Network, Peer-to-Peer network, Perceptive intelligence, Personal Data, Personal Data Sovereignty, Personal Information, Personal Information Right, Personality right, Personality trust, Physical currency, Point data, Peer-to-peer protocol, Polymerization effect, Popular sovereignty/sovereignty lies with the people, Positive human rights, Post-industrial Age, Power datamation, Precious Metal Credit, Prisoner's dilemma, Privacy, Private attribute of the right, Private digital currency, Private right, Production relations, Productivity, Programmable economy, Programmable finance, Programmable money, Programmable society, Property right, Public Data Rights System, Public Power, Public Power Governance, Public Power Property

Q Quantum computing, Quantum information, Quantum intelligence, Quantum mechanics, Quantum technology

R Rational order, Regenerative medicine, Reliability, Remixing, Rent-seeking, Right of Data self-defense, Right of use, Right to data equality, Right to data-generated remuneration, Right to dispose, Right to earnings, Right to know data, Right to use data, Rights of possession, Robot, Rule of China, Rule of conscience, Rule of Law Economy, Rule of system, Rule of technology

S Satoshi Nakamoto, Science and technology ethics, Self-activation, Self-empowerment, Self-processing, Sha Shu (taking advantage of acquaintances), Shadow Economy, Share, Sharing Culture, Sharing Economy, Sharing Right, Sharing System, Simplified mechanism, Singularity, Smart city, Smart collision, Smart contract, Smart family, Smart Internet, Smart manufacturing, Smart society, Smart terminal, Social capital, Social class/stratum, Social control, Social cooperation, Social generative theory, Social governance, Social integration, Social joint rights, Social mobilization, Social order, Social organization, Social rights, Social structure, Social trust theory, Sovereign attribute, Sovereign identity, Sovereignty authority, Sovereignty barrier, Sovereignty blockchain, Sovereignty cooperation, Sovereignty currency, Sovereignty demisability, Sovereignty digital currency, Sovereignty interest, Sovereignty Internet, Sovereignty transfer, Sovereignty will, Sovereignty of cyberspace, Space Battle, State Data Sovereignty, State Governance, State Sovereignty, Status Society, Stranger society, Stratification, Streamlining administration and delegating more power to lower level governments, Strip data, Super sovereign currency, Suspicions chain, System of usufructuary rights of data

T Targeted poverty alleviation, Tech for Social Good, Technical dilemma, Technical liberalism, Technocracy, Terahertz Technology, Ternary world, The Era of Combine of Labor, The extension of conscience, The fifth-generation mobile communication technology, The first generation of human rights, The fourth generation of human rights, The fourth industrial revolution, The Internet of Value, The right to freedom, The second generation of human rights, The third generation of human rightsm, Theory of smart contract, Theory of the value of sharing, Third-Party Payment, Time stamp, Traceability, Transaction book, Trust, Trust and dependence, Trust component theory model, Trust culture, Trust digitization, Trust Machine, Trust mechanism, Trust model, Trust system, Turing Test

U Ubiquitous Network, Unilateral Control, Unilateralism, Unity of knowledge and action, Universal equivalent

V Value delivery, Value exchange, Value network, Value nihilism, Value transfer, View of International Order, View of sovereignty, View on global governance, Virtual bullying, Virtual Reality (VR), Virtual society, Virtual space, Virtual violence, Virtual world

W Wang Yangming's Philosophy/Wang Yangming's Idealistic School

Y Yangming Philosophy

Z Zero marginal costs, Zero-sum game

Bibliography

I. Chinese Monographs

Chen Guoqiang. *A Concise Anthropological Dictionary of Culture*. Zhejiang People's Publishing House, 1990.

Key Laboratory of Big Data Strategy. *Block Data 5.0: Theory and Methodology of Data Sociology*. CITIC Press, 2016.

Key Laboratory of Big Data Strategy. *Data Rights Law 2.0: Institutional Construction of Data Rights*. Social Sciences Academic Press, 2018.

Li Kaifu, Wang Yonggang. *Artificial Intelligence*. Cultural Development Press, 2017.

United Nations Commission on Global Governance. *Our Global Neighborhood*. Oxford University Press, 1995.

Liu Pinxin. *Internet Law*. China Renmin University Press, 2009.

Research on Principles of Personal Data Protection Law and Legal Issues of Transnational Circulation. Wuhan University Press, 2004.

China Institute of Digital Assets. *Libra: An Experiment in Financial Innovation*. Oriental Press, 2019.

Su Li: *Law Book Review*. Law Press · China, 2003.

[German] Marx, [German] Engels. *Marx and Engels Collected Works (Volume 1), (Volume 46, Part 1)*, translated by the Central Compilation and Translation Bureau of the Works of Marx, Engels, Lenin and Stalin. People's Publishing House, 1979.

[US] Nassim Nicholas Taleb. *Antifragile: Things That Gain from Disorder*, translated by Yu Ke. CITIC Press, 2013.

[US] Piero Scaruffi, Niu Jinxia, Yan Jingli. *Humankind 2.0*, CITIC Press, 2017.

[UK] Adam Smith. *An Inquiry into the Nature and Causes of the Wealth of Nations*. The Commercial Press, 2011.

II. Chinese Journals

Chen Zhiyuan. Changed Bacteriological Warfare Based upon Chinese, Russian, American and Japanese Historical Materials. *Hunan Social Science*, 2016, No. 1.

Key Laboratory of Big Data Strategy. Ten Paths to Empower Social Governance through Blockchain. *Information for Deciders Magazine*, 2019, No. 47.

© Zhejiang University Press 2021
L. Yuming, *Sovereignty Blockchain 1.0*,
https://doi.org/10.1007/978-981-16-0757-8

Liu Shuchun. Strategic Implication, Technical Framework and Path Design of Digital Government Based on the Practice and Enlightenment of the "Zhejiang Pilot". *Chinese Public Administration*, 2018, No. 9.

Peng Yun. Research on authenticating data rights in Big Data environment. *Modern Science & Technology of Telecommunications*, 2016, No. 5.

China National Committee for Terms in Sciences and Technologies, Key Laboratory of Big Data Strategy. Top 10 New Terms for Big Data. *China Terminology*, 2017, No. 2.

Shen Mingming. Philosophy: The "Turn" of the Human Spiritual World: A Reconsideration of the Relationship between Science and Philosophy. *Fujian Tribune (The Humanities & Social Sciences Monthly)*, 2000, No. 6.

Xu Jing. Our Shared Digital Future: Building an Inclusive, Trustworthy and Sustainable Digital Society. *The Internet Economy*, 2019, No. 5.

Zhang Ming, Zheng Liansheng. The Subprime Mortgage Crisis Goes Deeper, Probing into the Crises of Fannie Mae and Freddie Mac. *Modern Bankers*, 2008, No. 8.

Zhao Jinxu, Meng Tianguang. Technology Empowerment: How the Blockchain Reshapes Governance Structures and Models. *Contemporary World and Socialism*, 2019, No. 3.

Zheng Gang. Financial Attacks: a New Form of Stealth Warfare. *Competitive Intelligence*, 2013, No. 3.

IV. Other Chinese References

Bian Zhe. Blockchain Technology—Building Data Trust for Digital Governance, *GMW.cn*. 2019. https://theory.gmw.cn/2019-11/04/content_33291595.html

Zhu Yan. Promoting the Construction of a Community with a Shared Future for Humanity through Blockchain and Other Technical Means. MBA china.com. 2018. http://www.mbachina.com/html/tsinghua/201809/168431.html

Printed in the United States
by Baker & Taylor Publisher Services